T0211068

Introduction to
PERCOLATION
THEORY

Revised Second Edition

Introduction to
PERCOLATION THEORY

Revised Second Edition

DIETRICH STAUFFER
Professor of Theoretical Physics
University of Cologne, France

AMNON AHARONY
Director, Sackler Institute of Solid State Physics
Tel Aviv University, Israel

CRC PRESS

Boca Raton   London   New York   Washington, D.C.

Library of Congress Cataloging-in-Publication Data

Catalog record is available from the Library of Congress

Visit the CRC Press Web site at www.crcpress.com

© 1991, 1994 by Dietrich Stauffer and Amnon Aharony

No claim to original U.S. Government works
International Standard Book Number 0-7484-0253-5
Printed in the United States of America   2 3 4 5 6 7 8 9 0
Printed on acid-free paper

Contents

Preface to the Second Edition

This edition is enlarged not only by a second author, but also by additional material, needed because of the progress in research since the first edition. In addition to correction of errors, omission of topics which are no longer active and updating of references, the new edition puts more emphasis on fractals and on the relations between the fractal geometry of percolation clusters and their physical properties. It also puts more emphasis on the crossover (as function of length scale) between the critical behaviour at and near the percolation threshold and other types of behaviour, and on the presentation of the renormalization group as a working tool. Dimensions above three are no longer ignored.

Most of the new material in this edition is contained in Chapters 5, 6 and 7, and in Appendix B. New concepts which are introduced, at least briefly, include distribution functions, fluctuations, multifractals, minimal (chemical) paths, continuum percolation, elastic networks, superlocalization, fractons, hulls and perimeters, diffusion fronts, growth, self-affine curves, invasion percolation and self-organized criticality. We also discuss the role played by percolation theory in describing Ising model droplets, leading to the very fast modern Monte Carlo simulation techniques in statistical physics. Many of the concepts listed here are not limited to percolation clusters. They arise in many modern branches of statistical physics and other disciplines which involve the interplay of geometry, statistics and physics. Examples include polymer science, aggregation, etc. Percolation theory is still the simplest context in which all these tools can be introduced and explained.

Another new element in this edition is the set of exercises. Many of these were originally research problems, and their solutions are hidden in the 'further reading' lists or in references therein. However, any student who follows the text should be able to solve these exercises, and use them to test his or her ability to actively use the tools offered here in research.

Obviously, this book still represents our biased view of percolation. One cannot do justice to all the topics which have appeared in the literature, and it is also beyond the scope of such an introductory text. Since this book is not a review monograph, we also avoided giving a full list of references. In many cases we have preferred to list a recent review or research article which contains many of the earlier references. In any case, the responsibility for these omissions, as well as for the remaining and new errors, rests fully with my coauthor.

Joan Adler, Preben Alstrom, Etienne Guyon, Shlomo Havlin, Rudolf Hilfer, Ury Naftaly and Gene Stanley made comments on the first version which were helpful in improving the book.

Dietrich Stauffer
Amnon Aharony
Tel Aviv, January 1991

Preface to the First Edition

This book is an attempt to introduce the reader to a research field which is already more than forty years old but which has become very fashionable in the research publications of the last ten years. More than a hundred publications are printed each year where 'percolation' or similar words appear in the title. But in contrast to many other modern research fronts, percolation theory is a problem which is, in principle, easy to define. It is, however, not so easy to solve. Thus percolation theory gives the reader the opportunity to enter current research without having to hear many specialized courses or to read voluminous textbooks. Percolation theory has been taught to first year undergraduates and even younger students by this author and by others, but it has also been utilized in courses for more advanced students on disordered systems or on computer simulations. The present book tries to be useful for all these purposes. Since percolation is not, and in my opinion should not be, a standard subject for university examinations, it is addressed mainly to readers interested in research. But it is aimed at one who is just starting research in this field, not to readers who have already worked in percolation theory. These experts will doubtless notice my biased selection of material.

The prerequisites are familiarity with computer programming (e.g. Fortran), integration and differentiation of functions of one variable, and probabilistic elements like averages or statistical independence. Therefore the book should be understandable to students in computer science, mathematics, chemistry or biology who might be interested in stretching a big computer to its limits by efficient programming, in simple applications of probability theory, in theories for the gelation of branched macromolecules, or in the spread of epidemics in an ensemble of living beings. But presumably the book will be read mostly by physics students since the methods by which it attacks the percolation problem are taken from the theory of phase transitions like ferromagnetic Curie points. For these we will mention throughout the book analogies between the geometrical aspects of percolation and the physical aspects of thermal phase transitions. Any modern textbook on Statistical Physics will give the background to understand these analogies, and students not interested in the analogies may simply ignore them.

Some readers may be acquainted with the scaling theory of phase transitions developed during the last twenty years and honoured by the 1982 Nobel prize for Kenneth G. Wilson. They will notice that many aspects of percolation theory are simply borrowed from the physics of phase transitions, as will be mentioned in the text. On the other hand, a reader who has not yet learned scaling theory or renormalization group methods for general phase transitions, but who wants to know something about them, may use percolation theory as a starting point. In many respects percolation is the simplest not exactly solved phase transition and thus may serve as an introduction to the sometimes more difficult review articles or books on phase transitions and critical phenomena.

What this book does not try to be is mathematically rigorous or complete in dealing with the actual state of research. We only give cursory mention to the applications of percolation since they require a more specialized readership. We will try however to list in the literature some of the more thorough review articles on percolation for those who want to study this field further. However, because of the rapid development of percolation, the reader should not assume that these references are still the most recent relevant reviews or original articles at the time he reads this book.

I am indebted to J. Kertész for information about his percolation seminar at Munich Technical University, and his comments and those of D. W. Heermann, H. J. Herrmann, A. Margolina, B. Mühlschlegel, R. B. Pandey, S. Redner, and M. Sahimi on a preliminary version of the manuscript (though I did not follow all of their suggestions, like calling this book 'My biased view of percolation'). M. Suessenbach for producing Figure 2, and A. Schneider for drawing the other figures. The manuscript was written using the text editing system of a PDP 11/34 computer, thanks to the efforts of A. Weinkauf and M. Schulte; needless to say this computer should be blamed for all the errors in the book.

Dietrich Stauffer
January 1985

CHAPTER 1

Introduction: Forest Fires, Fractal Oil Fields, and Diffusion

1.1. WHAT IS PERCOLATION?

Imagine a large array of squares as shown in Fig. 1(a). We imagine this array to be so large that any effects from its boundaries are negligible. Physicists call such an array a square lattice, mathematicians denote it by $\mathbb{Z}^2$; common sense identifies it with a big sheet of ruled paper. (You may complain that the square lattice in Fig. 1(a) is not very large, but the publisher did not allow us to fill all remaining pages of this book with these squares, which would have greatly simplified our task of writing the book and yours of reading it.) Now a certain fraction of squares are filled with a big dot in the centre, whereas the other squares are left empty, as in Fig. 1(b). We now define a *cluster* as a group of neighbour squares occupied by these big dots; these clusters are encircled in Fig. 1(c). From this picture we see that squares are called nearest neighbours if they have one side in common but not if they only touch at one corner. Physicists call squares with one common side 'nearest neighbour sites on the square lattice', whereas squares touching at one corner only are 'next nearest neighbours'. All sites within one cluster are thus connected to each other by one unbroken chain of nearest-neighbour links from one occupied square to a neighbour square also occupied by a big dot. The graphical 'cluster' explanation through Fig. 1(c) seems more appropriate for our purposes here than a precise mathematical definition. Percolation theory now deals with the number and properties of these clusters; perhaps the reader will agree with us that there are not many requisites needed to understand what percolation theory is about.

How are the dots distributed among the squares in Fig. 1? One may assume that the dots love to cling together, or that they hate each other and try to move as far away as possible. But the simplest assumption is that they ignore each other, not unlike scientists working in similar fields. Then the occupation of the squares is *random*, that is each square is occupied or empty independent of the occupation status of its neighbours. We call p the probability of a site being occupied by a big dot; that means that if we have N

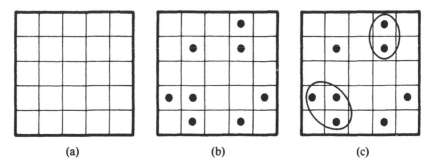

Fig. 1. Definition of percolation and its clusters: (a) shows parts of a square lattice; in (b) some squares are occupied with big dots; in (c) the 'clusters', groups of neighbouring occupied squares, are encircled except when the 'cluster' consists of a single square.

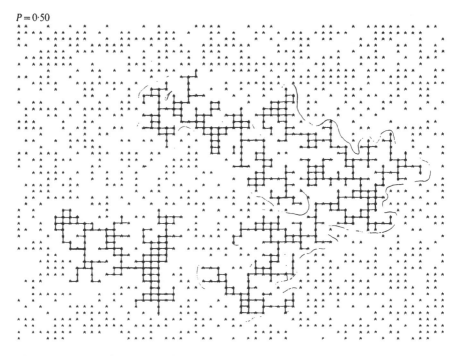

Fig. 2. Example for percolation on a 60×50 square lattice, for various p as indicated. Occupied squares are shown as *, empty squares are ignored. Near the threshold concentration $0 \cdot 5928$ the largest cluster is marked.

P = 0·60

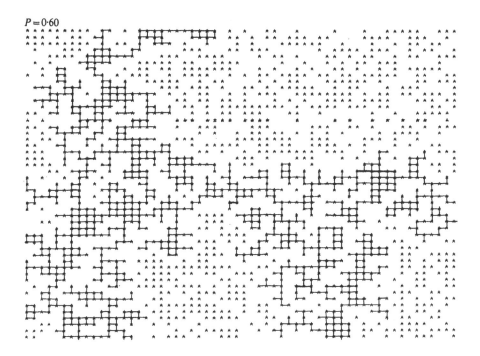

P = 0·70

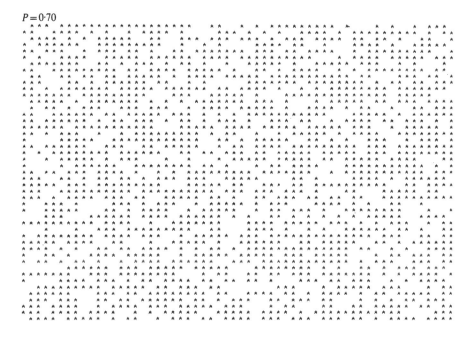

squares, and N is a very large number, then pN of these squares are occupied, and the remaining $(1 - p)N$ of these squares are empty. This case of random percolation is what we concentrate on here:

Each site of a very large lattice is occupied randomly with probability p, independent of its neighbours. Percolation theory deals with the clusters thus formed, in other words with the groups of neighbouring occupied sites.

Of course, the reader may replace 'occupied by a big dot' with 'black' and 'empty' with 'white' (or 'red', if he likes politics); or he may use any other suitable pair of words denoting two mutually exclusive states of the site.

Figure 2 shows a computer-generated sample of a 60×50 square lattice, with probability p increasing from 50% to 70%. We see that for $p \geqslant 0 \cdot 6$ one cluster extends from top to bottom and from left to right of the sample; one says that this cluster percolates through the system rather like water percolates through a coffee machine. A large part of this book deals with the peculiar phenomena of percolation near that concentration p_c where for the first time a percolating cluster is formed. These aspects are called *critical phenomena*, and the theory attempting to describe them is the *scaling theory*.

Historically, percolation theory goes back to Flory and to Stockmayer who during World War II used it to describe how small branching molecules form larger and larger macromolecules if more and more chemical bonds are formed between the original molecules. This polymerization process may lead to *gelation*, that is to the formation of a network of chemical bonds spanning the whole system. Thus the original small molecules correspond to our squares, the macromolecules to our clusters, and the network to our percolating cluster. You may be an experienced researcher in percolation without having been aware of it, for the boiling of an egg, which is first liquid and then becomes more solid-like ('gel') upon heating is an example. Flory and Stockmayer developed a theory which today one calls percolation theory on the Bethe lattice (or Cayley tree) and which will be explained later. But until recently it was controversial whether critical phenomena for gelation are described correctly by percolation theory and its assumption that chemical bonds are formed randomly (de Gennes, 1976; Kolb and Axelos, 1990).

Usually, the start of percolation theory is associated with a 1957 publication of Broadbent and Hammersley which introduced the name and dealt with it more mathematically, using the geometrical and probabilistic concepts explained above. Hammersley, in his personal history of percolation in *Percolation Structures and Processes*, mentions that the new computers which became available to scientists at that time were one of the reasons for developing percolation theory as a problem where the computers could be useful. We will see later that even today computers play a crucial role for percolation, with lattices containing thousands of millions of sites being simulated and analysed.

The percolation theory as described here, with its particular emphasis on critical phenomena, was developed since the 1970s; one may regard a note by Essam and Gwilym in 1971 as one of the starting points of the later avalanche of publications. Instead of going through the details now we describe three simple 'games' which can be easily simulated on a computer and which may serve as an introduction to a reader preferring to learn percolation by a 'hands-on' approach. These example are somewhat unusual, and the reader may skip them and proceed with Chapter 2.

1.2. FOREST FIRES

This section introduces a simple model for forest fires. Its aim is not so much to help fighting fires but to help to understand the idea of a percolation threshold, the concept of a sharp transition with diverging times, and computer simulation.

French scientists in Marseilles and elsewhere are interested, for obvious reasons, in understanding and controlling forest fires. They told us of the following percolation problem which can easily be simulated on a computer. How long does a forest fire take to either penetrate the forest or to be extinguished?

As is well known, a diligent student should make hundreds of independent experiments to reduce statistical errors before reporting the results in his thesis. If for every thesis, a hundred fires were initiated in the forests surrounding the university, society's respect for research might be diminished. It is much more practical to simulate numerous such fires on a computer. For this purpose we approximate the forest by a square lattice. Each square in Fig. 1 is either occupied by a tree, in which case we call that site 'green', or it is empty, in which case we call it 'white'. The probability for a green square is p, that for a white square is $(1 - p)$. For $p = 1$ all squares would correspond to trees, which would be appropriate to a garden of apple trees but not for a natural forest. The fact that $p < 1$ allows for holes (white squares) which cause disorder in the forest. This distribution of white and green sites (squares) is our initial state.

Now let some trees burn and call those squares which correspond to burning trees 'red' sites. The simplest choice is to light all the trees in the first row of the lattice, whereas the remaining trees, in lines 2, 3, ..., L of the $L \times L$ lattice, remain green. Does this fire on one side of the forest penetrate through the whole forest down to line L of our array?

For this purpose we have to clarify how a tree can ignite the other trees. To simplify the computer simulation we go through our lattice regularly, first scanning the first line of trees from left to right and checking which neighbours they ignite, then scanning the second line in the same way, and so on until we reach the last line of trees. During the whole simulation, a green tree is ignited and becomes red if it neighbours another red tree which at that time is still burning. Thus a just-ignited tree ignites its right and bottom neighbour within the same sweep through the lattice, its top and left neighbour tree at the next sweep. Reaching the end, we start again with the tree at the extreme left in the first line. Each sweep through the whole lattice (experts call that one Monte Carlo step per site) constitutes one time unit in our simulation. We assume that the fire can spread only to green nearest neighbour trees, not to trees which are farther away. Furthermore, a tree which has burnt during one time unit is regarded as burnt out ('black') and no longer ignites any other tree. We regard the forest fire as terminated if it either has reached the last line or if no burning trees are left. (In the first case, the fire would ignite the next line of trees if a larger lattice had been stored in the computer; in the second case, only black trees and green trees adjoining white places are left over, the black trees constituting formerly burning trees which have burnt out, the green trees never having been touched by the fire since they were separated safely from the burning trees). The *lifetime* of the forest fire is defined as the number of sweeps through the lattice until termination is reached, averaged over many distributions of trees among the sites of the same lattice at the same probability p.

Figure 3 shows this lifetime of forest fires as a function of the probability p that a square is occupied with a tree. These simple computer simulations indicate that there is a sharp transition, for the above case near $p = 0 \cdot 6$, where the lifetime seems to approach infinity. Of course, in the simulation of finite lattices the reader cannot expect truly infinite times; but one can simulate the forest fires at the same 'critical' value of p near $0 \cdot 5928$ for different lattice sizes and show that the lifetime increases with increasing size of the forest.

Why is there a special value of p, which we call the percolation threshold p_c, where the lifetime seems to diverge? For p near unity, each row can imme-

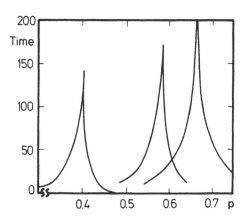

Fig. 3.   Average termination time for forest fires, as simulated on a square lattice. The centre curve corresponds to the simplest case described in the text. The lefthand curve gives data if the fire can spread to both nearest and next nearest neighbours. For the righthand curve two burning trees are needed to ignite a nearest or next-nearest neighbour.

diately ignite the trees in the next row, and thus after one sweep through the lattice the fire may already have reached the last row. For p near zero, most burning trees have no neighbours at all, and the fire stops there after the tree has burnt out; thus after a few sweeps nothing burns anymore. If we increase p from small values to large values, then at some critical value $p = p_c$ a path of neighbouring trees appears which connects the top row with the bottom row for the first time, that is we see a percolating cluster. The shortest path which, for p slightly above p_c, this percolating network creates to connect top and bottom is called the minimal, or chemical path. It will in general be very different from a straight line. Fig. 4 shows a typical path. (See also Section 5.2.)

Because of the simplified way in which we construct our model, the fire spreads preferentially from top to bottom, or left to right, and needs a much longer time to move backwards from right to left or from bottom to top. For four consecutive forward steps, say top to bottom, it needs just one time unit, whereas four backward steps require four time units, as the reader can easily check on this figure by going through the above algorithm. Thus now the fire needs a long time to penetrate the forest. If p is diminished to a value slightly below p_c, then some trees, for example the one marked by an X in Fig. 4, may be missing. The fire then needs a long time to find out that it cannot penetrate the forest, and thus only after many sweeps through the lattice will the fire be extinguished. Therefore the lifetime will become very large if p approaches p_c from below or above.

We also show in Fig. 3 the results for two modifications of the above model. In one case we allow the fire to spread not only to the nearest neighbour trees (squares which have one side in common) but also to next-nearest neighbours (squares which have only one corner in common). Then the critical

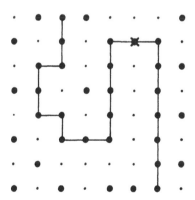

Fig. 4. Example of the shortest path connecting the top line of a small square lattice with the bottom line, for p slightly above p_c. The straight sections of this line connect the centres of occupied squares, the X marks the site which, if missing due to a small reduction of p, would disconnect top and bottom lines but would still give a long termination time for the forest fire simulation.

point is shifted to about $0 \cdot 4$; experts have shown that it is at one minus the above critical value, i.e. $1 - 0 \cdot 5928 = 0 \cdot 4072$. But even without much thinking and computing one can understand that now the fire can spread more easily since it can jump over longer distances; therefore the percolation threshold is lowered.

The other modification goes in the opposite direction: We assume that the weather is more like in Nova Scotia (Canada) than in Marseilles (France). Since it is quite cold, a tree needs two burning neighbours, instead of only one, before it can ignite. Now it is more difficult for the fire to percolate through the forest, and the percolation threshold is shifted upwards, as the simulations in Figure 3 show.

The reader may complain that the above algorithm gives the fire a preference to spread to the right and the bottom and may dislike these similarities with political or economic trends, respectively. But for forest fires, such trends can be justified as representing a wind blowing in one 'diagonal' direction. In reality this preference is introduced to save computer time.

For readers interested in the physics of phase transitions it should be mentioned that the percolation threshold at $p = p_c$ gives the position of a phase transition (for experts only: without 'broken symmetry'). At a phase transition, a system changes its behaviour qualitatively for one particular value of a continuously varying parameter. In the percolation case, if p increases smoothly from zero to unity, then we have no percolating cluster for $p < p_c$ and (at least) one percolating cluster for $p > p_c$. Thus at $p = p_c$, and only there, something peculiar happens: for the first time a path of neighbouring green trees connects top and bottom. Also the divergence of characteristic times (in our case the fire lifetime) at the critical point has analogies in other phase transitions where it is called *'critical slowing down'*. For example, for a temperature only slightly below the liquid–gas critical temperature, the fluid is quite unsure whether it wants to be liquid or vapour, and thus takes a lot of time to make its choice; this time can be measured by light scattering. Similarly, relaxation times near magnetic Curie points are very large.

1.3. OIL FIELDS AND FRACTALS

Percolation can be used as an idealized simple model for the distribution of oil or gas inside porous rocks in oil reservoirs. In Fig. 1, imagine that the unoccupied (white) squares represent regions filled with hard rock, while the occupied squares represent pores that are filled with oil or gas. The average concentration of oil in the rock is represented by the occupation probability p. (In the oil terminology, p is called 'porosity'.) In real reservoirs, the mechanisms that created the oil deposits imply some *correlations* between occupied pores, owing to the way the rock was originally cracked or the way the different deposits were put in place. The simple percolation model ignores these correlations, and assumes that each basic square (or cube, if this is repeated in three dimensions) is occupied or empty independently of its neighbours. However, the qualitative features described below also hold in the more realistic models.

It is obvious from Fig. 2, that when p is smaller than p_c, the oil is found only in finite connected clusters. Therefore if we place a well at a random site, it will most probably hit a small cluster, produce a finite small amount of oil and be a very bad investment. To produce a large amount of oil, we need a reservoir which has $p > p_c$, and we need to have the well at a site that belongs to the largest cluster.

The oil people are very interested in predicting how much oil they would produce from a well. To help in these predictions, they take out rock samples from the well. These come in long rock logs, with a typical diameter of order 5–10 cm. One can then measure the porosity (percentage of pores) in a piece of linear size 5 cm, and try to *extrapolate* to the reservoir scale, which could be many kilometres. Is such extrapolation legitimate?

To address this question, let us identify our 'well' as one site sitting in the square example of Fig. 2 and belonging to the largest cluster for $p > p_c$. Let us next put a frame of size $L \times L$ around this point, and count how many points within this frame belong to the same cluster, $M(L)$. The reader can easily try this exercise with frame sizes $L = 3, 5, 7, 9$, etc. Looking at the last example in Fig. 2, it is clear that $M(L)$ practically grows linearly with the area of the frame, L^2, and we can define the average density of points connected to our well as $P = M(L)/L^2$. P is then independent of L, and is monotonically decreasing as p decreases. However, the situation is very different for p very close to p_c, e.g. $p = 0 \cdot 6$. In that case, the largest cluster is rather *ramified*, and it has many 'holes' in it. Those holes contain other clusters, which may be quite large, but whose oil is not reachable through our well. Looking at the picture for $p = 0 \cdot 6$ in Fig. 2, one sees 'holes' *on many length scales*. As we shall see later, the occurrence of phenomena on all length scales is very basic for many of the interesting phenomena, which occur near p_c.

If one measures $M(L)$ as function of L at p_c, the result is no longer linear in the area L^2. In fact, if one plots $\log M(L)$ versus $\log L$ for bigger lattices (see also Fig. 15), one finds a straight line with slope $1 \cdot 9$, implying that at p_c one has

$$M(L) \propto L^{1 \cdot 9}$$

(We use the symbol $\propto$ to indicate proportionality. In many cases this proportionality is meant to be accurate only in the *asymptotic* limit, here of large L.) The exponent $1 \cdot 9$ is called a 'fractal dimensionality', or fractal dimension.

You may have noticed with sadness that a small bottle of scotch, half as high as the customary whisky bottle, does not contain half as much of the precious fluid but only one eighth; not only the height is reduced by a factor 2 but also the width and the depth, with the volume being the product of these three lengths. In other words, a bottle is a three-dimensional object. For two dimensions, a piece of paper which has a length and a width both twice as large as those of another piece weighs four times as much. Only a one-dimensional object, like a long wire, is simple. A wire half the length of another weighs half as much as the longer wire. In all these cases, the mass M scales with the linear size L as $M \propto L^d$, and d is the usual *Euclidean dimension*. Benoit Mandelbrot introduced 'fractal geometry' as a unifying description of natural phenomena which are not uniform but still obey simple power laws of the form

$$M \propto L^D$$

with non-integer dimensions D. For three-dimensional percolation clusters at p_c one finds $D \simeq 2 \cdot 5$. We shall see more examples of *fractals* below.

The fact that $M(L)$ grows as $L^{1 \cdot 9}$ implies that the average density $M(L)/L^2$ is not constant, but rather decays as $L^{-0 \cdot 1}$. Therefore, the average density of the extractable oil in a field with porosity near p_c, of size 100 km, is smaller by a factor of about $(10^6)^{-0 \cdot 1} \simeq 0 \cdot 25$ than that of a sample of size 10 cm. The remaining 75 per cent is not directly connected to the drill hole. Such a factor is crucial if we are to base the economy of oil production on it! The corresponding factor in three dimensions is $(10^6)^{-0 \cdot 5} = 10^{-3}$!

In fact, the situation is not so bad, since the density does become uniform for large L above p_c. As we shall see, there exists a typical length $\xi(p)$, called the *correlation length*, such that $M(L) \propto L^{1 \cdot 9}$ for $L < \xi$, and $M(L) \propto L^2$ for $L > \xi$. ξ is a measure of the largest hole in the largest cluster, and it decreases as we increase p above p_c. However, the oil people should use a sample larger than ξ in order to estimate the correct amount of oil they can get. A more quantitative discussion of this problem will be given below, in Section 3.4.

The problem of extracting oil from the rock involves not only estimating the amount of such oil, but also discussing the *flow* of the fluid inside the porous medium. This brings up many questions concerning *dynamics* on the percolation clusters, that we shall discuss below. The simplest example, concerning *diffusion*, is briefly introduced in the next section.

The reader should be warned, however, that both these remarks on oil flow, as well as the earlier ones on forest fires, are meant as illustrations, not as proven engineering applications.

1.4. DIFFUSION IN DISORDERED MEDIA

Hydrogen atoms are known to diffuse through many solids, an effect which might become important for energy storage. If the solid is not a regular lattice, this diffusion takes place in a disordered, not an ordered medium. A particularly simple disordered medium is our percolation lattice, where only a fraction p of all sites (squares) is occupied, the rest are empty. Let us assume the hydrogen atom can move only from one occupied site of the lattice to a nearest neighbour which is also occupied. Then the motion is restricted to the cluster of percolation theory to which the atom belongs initially. It can never jump to another cluster since then it would have to move at least once over a distance larger than that between nearest neighbours. This problem was called the '*ant in the labyrinth*' by de Gennes in 1976. At the beginning of the 1980s this problem became very fashionable, particularly at the percolation threshold $p = p_c$.

Another useful application of this would concern the diffusion of test particles through the oil in the porous rock, mentioned above. Such diffusion is sometimes used to study the properties of the pore structure.

Let us not care whether hydrogen atoms move through solids, an ant tries to escape a labyrinth, or the reader desperately searches for a way through this book. We simply have a point, called an ant, which sits on an occupied square of our square lattice and which at every time unit makes one attempt to move. This attempt consists in randomly selecting one of its four neighbour squares. If that square is occupied, it moves to that square; if instead it is empty, the ant stays at its old place. In both cases the time t is increased by one unit after the attempt. After a certain time t, one calculates the squared distance between the starting point and the end point. One repeats the simulation by giving the ant a different occupied square as a starting point; finally, one averages the squared distance obtained in this way over many ant movements on many lattices at the same p and same lattice size. How does R, the square root of this averaged squared distance (also called the *root mean square* or rms displacement) depend on time t?

For $p = 1$ one has diffusion on a regular lattice without disorder, and elementary statistical considerations give $R^2 = t$ exactly, if our squares have a length equal to unity. (Proof: For each such random walk, the end-to-end vector R is the vector sum of t displacement vectors d_i, $i = 1, 2, ..., t$. When we calculate the square of that sum and then its average, we have to calculate the averages of the scalar products $d_i d_j$. For $i = j$, this scalar product is simply the square of the nearest neighbour distance, which is unity. For i and j different, the scalar product can be $+1$ or -1 with equal probability since we assumed that the motion is completely random. Moreover, in half of the cases the scalar product is zero since d_i and d_j are perpendicular to each other. Thus on average this product cancels out except for $i = j$ where it gives unity. Therefore the squared sum equals t. This proof is not necessary to understand the remainder of the book since we will mainly deal with problems which are not exactly solved.)

Figure 5 shows the results of simple computer simulations on the square lattice. On this double-logarithmic plot one sees the power law $R = \text{const} \times t^k$ more easily than on a normal diagram. It seems to describe the relation between distance R and time t for sufficiently large t. Since $\log R = \log (\text{const}) + k \log (t)$, power laws show up as straight lines in such log-log plots, with the slope giving the exponent k of the power law. We see that for a concentration p far above $p_c = 0 \cdot 59$, k is near $1/2$ for large times, whereas for p far below p_c the distance R approaches a constant for large

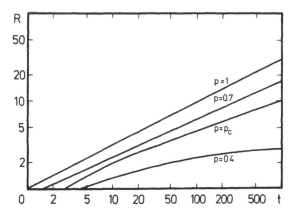

Fig. 5. Example of the distance R travelled by an ant in a labyrinth, as simulated on a square lattice for $p < p_c$, $p = p_c = 0 \cdot 5928$, and $p > p_c$. Note the double-logarithmic scales.

times, that is $k = 0$. Right at p_c, k takes on a value in between these two extremes and is roughly equal to $1/3$. This effect, that the exponent is neither that for normal diffusion ($k = 1/2$) nor that for a constant distance ($k = 0$) was called anomalous diffusion by Gefen, Aharony and Alexander. Again it has an analogue in critical points near thermal phase transitions. For example, spin diffusion in ferromagnets at $T = T_c$ or mass diffusion at liquid–gas critical points no longer follows the normal diffusion laws but is often described by an anomalous diffusion exponent k or $1/z$.

It is easy to understand why the ant moves so differently for p above and below the percolation threshold p_c. For $p < p_c$ there are only finite clusters present, and the ant sits on one of them. Thus it moves only within that cluster (if this cluster happens to be an isolated occupied square, the ant cannot move at all.) Therefore its motion is restricted over finite distances, and R approaches a value connected to the cluster radius if t is very large. For $p > p_c$, on the other hand, the ant can move to infinity if it starts on the percolating network. There are certain holes in this network; but for distances larger than the typical hole size, the ant feels only an average over the small holes, just as the tyres of your Rolls-Royce average over the small pores of the asphalt over which your chauffeur is driving you. Thus the disorder acts as a friction which slows down the diffusion process but does not prevent it: $k = 1/2$ for long times. Only at the border case $p = p_c$, does the ant not know which of the two power laws it should follow.

Considering t as the number of steps the ant performs, and R as the linear size of the region visited by the ant, the relation $t \propto R^{1/k}$ can also be interpreted as stating that the number of steps in a region of linear size R is *fractal*, with a fractal dimension equal to $1/k$. For regular lattices, and in the homogeneous regime describing the largest percolation cluster on large length scales above p_c, this exponent $1/k$ is equal to 2. At p_c, the ant is restricted to move on clusters which are themselves fractal. It is thus forced to move back and forth within a small piece of the cluster, until it finds its way out. This takes a long time and therefore the number of steps within a restricted area is large, and the fractal dimension of the walk, $1/k$, is larger than 2.

At intermediate concentrations, like $p = 0 \cdot 7$, the ant feels some fractal structure on short distances $(R < \xi)$. On those distances, the slope k is close to its value at p_c, i.e. $k \simeq 1/3$. Only when $R \gg \xi$ does the slope approach the uniform value $k = 1/2$. As seen from Fig. 5, the curve for $p = 0 \cdot 7$ has not yet become completely parallel to that of $p = 1$, even at $t = 500$. Better quality data, on larger samples, are needed to confirm the details of the crossover from anomalous to normal diffusion. The low-quality data of Fig. 5 are mainly meant to exhibit results that a student can readily produce on a personal computer. We shall present a more quantitative discussion of diffusion in Chapter 6.

1.5. COMING ATTRACTIONS

This introduction should have given you an impression of modern percolation theory. If you want to learn more, then Chapter 2 gives you a very detailed heuristic derivation, based on the exact solutions in one dimension and on Bethe lattices, of the crucial scaling assumption, Eq. (33), for the average number $n_s(p)$ of clusters, containing s sites each. The first and second moments of this cluster size distribution then give the 'strength' of the infinite cluster (critical exponent β) and the mean cluster size (critical exponent γ). The fractal geometry of clusters is the main content of Chapter 3. Finite size scaling and renormalization group methods are explained in Chapter 4. Chapter 5 deals with conductivity and multifractality and Chapter 6 discusses diffusion and related dynamic phenomena. The connection with thermal critical phenomena is discussed in Chapter 7. Appendix A deals with some numerical aspects, and Appendix B describes some dimension-dependent approximations.

FURTHER READING

General reviews are collected in the book *Percolation Structures and Processes*, edited by G. Deutscher, R. Zallen and J. Adler, published in 1983 by Adam Hilger, Bristol, as Annals of the Israel Physical Society 5. Out of the 21 articles there, our introduction is related in particular to that of Hammersley on the origins of percolation theory, of Jouhier *et al.* on gelation of macromolecules, and of Mitescu and Roussenq on ant diffusion.

For a more recent review, see Aharony, A. in: *Directions in Condensed Matter Physics*, edited by Grinstein, G. and Mazenko, G. (Singapore: World Scientific, 1986).

Mathematical aspects, mostly ignored by us, are emphasized by H. Kesten's book *Percolation Theory for Mathematicians*, (Boston: Birkhauser, 1982).

Other percolation reviews
Essam, J.W., *Rep. Prog. Phys.*, **43**, 843 (1980).
Guyon, E., Roux, S., Hansen, A., Bideau, D., Troadec, J.P. and Crapo, H., *Rep. Prog. Phys.*, **53**, 373 (1990).
Kirkpatrick, S., *Rev. Mod. Phys.*, **45**, 574 (1973).
Sahimi, M., *Applications of Percolation Theory* (London: Taylor & Francis, 1994).
Stauffer, D., *Phys. Rep.*, **54**, 1 (1979).
Physica A, **157**, 1–644 (1989). [Conference, ETOPIM II.]
Physica A, **168**, 1–676 (1990). [Conference, Domb's birthday.]
Proc. Roy. Soc., **423**, 1–200 (1989). [Conference, Fractals in the Natural Sciences.]

Books on disordered systems
Boccara, N. and Daoud, M., *Physics of Finely Divided Matter* (Heidelberg: Springer-Verlag, 1985).
Bunde, A. and Havlin, S., *Fractals and Disordered Systems* (Berlin: Springer, 1991).
Efros, A.L., *Physics and Geometry of Disorder* (Moscow: Mir, 1986).
Meakin, P. in: Domb and Lebowitz (1988) (see below), vol. 12, p. 336.
Pynn, R. and Riste, T. (editors), *Time Dependent Effects in Disordered Materials* (New York: Plenum Press, 1987).
Pynn, R. and Skjeltrop, A. (editors), *Scaling Phenomena in Disordered Systems* (New York: Plenum Press, 1985).
Stanley, H.E. and Ostrowsky, N. (editors), *On Growth and Form* (Dordrecht: Martinus Nijhoff, 1986).
Stanley, H.E. and Ostrowsky, N. (editors), *Correlations and Connectivity* (Dordrecht: Kluwer, 1990).
Zallen, R., *The Physics of Amorphous Solids* (New York: Wiley, 1983).

Fractals and diffusion
Avnir, D. (editor), *The Fractal Approach to Heterogeneous Chemistry* (New York: Wiley, 1989).
Feder, J., *Fractals* (New York: Plenum Press, 1988).
Guyon, E. and Stanley, H.E., *Fractal Forms* (Amsterdam: Elsevier, 1991) [photo album].
Havlin, S. and Ben Avraham, D., *Adv. Phys.* **36**, 695 (1987).
Mandelbrot, B.B., *The Fractal Geometry of Nature* (San Francisco: Freeman, 1982).
Vicsek, T., *Fractal Growth Phenomena* (Singapore: World Scientific, 1989).
*Physica D., * **38**, 1–398 (1989). [Conference, Mandelbrot's birthday].
*Physica A., * **191**, 1–577 (1992). [Conference on Fractals and Disordered Systems].

First percolation theory for polymer gelation
Flory, P.J., *J. Am. Chem. Soc.*, **63**, 3083, 3091, 3906 (1941).
Stockmayer, W.H., *J. Chem. Phys.* **11**, 45 (1943).

Phase transitions in general
Stanley, H.E., *Introduction to Phase Transitions and Critical Phenomena* (Oxford: OUP, 1971).
or the continuing series of books
Phase Transitions and Critical Phenomena, edited by Domb, C. and Green, M.S., then by Domb, C. and Lebowitz, J.L. (New York: Academic Press, 1972 onwards).

Percolation in semiconductors
Shklovskii, B.I. and Efros, A.L., *Electronic Properties of Doped Semiconductors*, (Heidelberg: Springer Verlag, 1984).

CHAPTER 2
Cluster Numbers

Percolation is a *random* process. Therefore, different percolation lattices will contain clusters of different sizes and shapes. In order to discuss their average properties, one must study the *statistics* of these clusters. This is done by studying the number of clusters with s sites per lattice site, $n_s(p)$. We start this chapter with exact calculations of these *cluster numbers* for the simple cases of one dimension, lattice animals and on Bethe lattices. These exact solutions first suggest a simple functional form, and this is then generalized to the scaling form of Eq. (33). The moments of the cluster numbers yield the strength P of the infinite cluster and the mean cluster size S, and scaling implies relations among the corresponding critical exponents and amplitudes. Series expansions and Monte Carlo simulations are used to check these theoretical predictions.

2.1. THE TRUTH ABOUT PERCOLATION

We did not tell you the whole truth: life is more than just a square lattice. There are also the triangular lattice, the honeycomb lattice, and other two-dimensional structures. In three dimensions we have the *simple cubic lattice*, the body-centred cubic (bcc) lattice, the face-centred cubic (fcc) lattice, the diamond lattice, among others. Dimensions higher than three are also useful to test theories, and usually are treated by the *hypercubic lattice*. In Fig. 1 we defined the square lattice through the centres of the squares shown there. We could also have defined it equivalently through the points where the lines in Fig. 1 cross.

Now in Fig. 6(a) the situation is different. When we put the sites of the lattice on the crossing points of the lines of Fig. 6(a) we obtain the *triangular lattice*; if instead we put them in the centres of the triangles with equal distance from the surrounding lines, we get the *honeycomb lattice*. (We do not recommend the use of the term hexagonal lattice.) Figure 6(b) consists of cubes and is called $\mathbb{Z}^3$ by mathematicians; it does not matter whether we put the lattice sites in the centres or on the corners of the cube. In a FORTRAN computer program one could store the sites of the simple cubic lattice in an array $A(i, j, k)$ whose indices i, j and k vary independently from 1 to L, where L is a large integer. The sites of the bcc lattice are both the corners and the

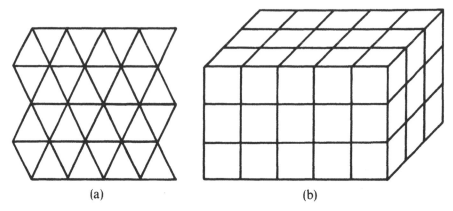

(a)   (b)

Fig. 6. Definition of triangular, honeycomb and cubic lattices. For the triangular lattice, every intersection of the lines in (a) is a lattice site; for the honeycomb lattice, the centres of the triangles in (a) form the lattice sites. The simple cubic lattice consists of the corners of the cubes in (b); for the bcc lattice the centres of the cubes, and for the fcc lattice the centres of the six faces of each cube, are added to the simple cubic lattice.

centres of the cubes, whereas the fcc lattice consists of the corners of the cubes and the centres of the six faces surrounding each cube. Diamond lattices are not the programmer's best friends. A five-dimensional hypercubic lattice is much easier to program, for example by using a FORTRAN array $A(i, j, k, m, n)$ with five independent natural numbers as indices (or different FORTRAN statements to the same effect) to simulate a part of $\mathbb{Z}^5$.

To demonstrate percolation experimentally on a triangular lattice we may put numerous small spheres of equal size but with two different colours (black and white, for example) into a large box. These balls will roll around on the bottom of the box if the box is large enough to prevent the spheres being on top of each other. Now if the box is slightly inclined all the balls will roll to one side of the bottom plane. By shaking the box slightly, the balls are persuaded to form a triangular lattice with a few defects. The two different colours then symbolize occupied and empty sites, and one sees the clusters quite directly. With an equal number of black and white balls, one is studying the behaviour at the percolation threshold $p_c = 1/2$ (see below). If the black balls are electrically conducting and the white balls are insulating, one can measure percolation electrically; but as the Marseilles group of Clerc *et al.* found out, such experiments are more difficult than the visual inspection recommended here.

For all these lattices, each site is randomly occupied with probability p and empty with probability $(1 - p)$ and clusters are groups of neighbouring occupied sites.

Everything we have defined so far is called '*site percolation*'. Its counterpart is called '*bond percolation*' and is defined as follows. Imagine each site of the lattice to be occupied, and lines drawn between neighbouring lattice

Table 1. Selected percolation thresholds for various lattices. 'Site' refers to site percolation and 'bond' to bond percolation. In all cases, only nearest neighbours form clusters, and no correlations are allowed between different sites or bonds. If the result is not exact (see text), the error probably affects only the last decimal.

Lattice	Site	Bond
Honeycomb	0·6962	0·65271
Square	0·592746	0·50000
Triangular	0·500000	0·34729
Diamond	0·43	0·388
Simple cubic	0·3116	0·2488
BCC	0·246	0·1803
FCC	0·198	0·119
$d = 4$ hypercubic	0·197	0·1601
$d = 5$ hypercubic	0·141	0·1182
$d = 6$ hypercubic	0·107	0·0942
$d = 7$ hypercubic	0·089	0·0787

sites. Then each line can be an open bond with probability p, or a closed bond with probability $(1 - p)$. (Simply identify yourself with a water molecule in a coffee percolator, or an oil molecule in a porous rock; you then can only flow through the open channels, not through the closed ones.) A cluster is then a group of sites connected by open bonds. When measuring the size of a cluster one has to define whether one counts the site content or the bond content. For example, are two sites which are connected by an open bond with each other and by closed bonds with all other sites called a cluster of size two (site) or of size one (bond)? Because of this ambiguity this book deals mostly with site percolation, even though bond percolation historically came first.

The *percolation threshold p_c*, Table 1, is that concentration p at which an infinite network appears in an infinite lattice. For all $p > p_c$ one has a cluster extending from one side of the system to the other, whereas for all $p < p_c$ no such infinite cluster exists. In finite systems as simulated on a computer one does not have in general a sharply defined threshold; any effective threshold values obtained numerically or experimentally need to be extrapolated carefully to infinite system size (called the thermodynamic limit by physicists accustomed to thermodynamics). If one has a mathematically exact (or at least plausible) calculation for p_c, then of course no such extrapolations are needed. Moreover, such exact results can be tested to check the reliability of numerical methods.

In Chapter 4 we will deal in greater detail with computer simulations to determine p_c accurately. 'Series' extrapolations will be explained in Section 2.7. Mathematical methods to calculate this threshold exactly are restricted so far to two dimensions, consistent with the experience in the field of phase transitions that three-dimensional problems in general cannot be solved exactly. The review of Essam (1972, in vol. 2 of Domb and Green; refer to introduction), as well as Kesten (1982), explain how two-dimensional thresholds can be derived mathematically for many simple lattices. But progress in this field is not easy. It took about two decades from the first numerical estimates in 1960 for square bond percolation, over non-rigorous arguments that $p_c = 1/2$ exactly, to a mathematical proof. But today we also know $p_c = 1/2$ for the triangular site, $p_c = 2 \sin (\pi/18)$ for the triangular bond, and $p_c = 1 - 2 \sin (\pi/18)$ for honeycomb bond percolation. For the honeycomb site problem, p_c seems to be smaller than $1/\sqrt{2}$; and for square site percolation, no plausible guess for a possibly exact result is known to us at present.

For *Bethe lattices* (Cayley trees), where every site has z nearest neighbours and there are no closed loops, we show below (Section 2.4) that $p_c = 1/(z - 1)$. In hypercubic lattices in d dimensions, each site has $2d$ nearest neighbours (e.g., 4 and 6 nearest neighbours on the square and simple cubic lattices). As we mention in Section 2.4, when d is very large then loops become irrelevant, and the behaviour on hypercubic lattices approaches that on Bethe lattices, hence $p_c = 1/(2d - 1)$. Table 1 summarizes exactly and approximately known percolation thresholds, for both site and bond percolation. Note that although the values for site percolation on the square and the simple cubic lattice seem to suggest the rule $p_c = 1/(2d - 2)$, this is definitely wrong at high d, where $p_c = 1/(2d - 1)$. One should thus be careful when one tries to conjecture general formulae based on a few numbers! However, we shall see that looking at results for different dimensions d does help to identify trends, and gives hints on the underlying mechanisms and structures.

In all the examples of the table, clusters are defined as groups of nearest neighbours which are occupied or connected by open bonds. One may also allow next-nearest neighbours to form clusters. Then in Fig. 1 not only the squares which have one side in common, but also those touching each other only at a corner, belong to one cluster if they are occupied. Even connections over longer distances have been introduced. The percolation thresholds then go to zero if this connection range goes to infinity. In this case, p_c also approaches zero as $1/(z - 1)$ where z is the number of 'neighbours' connected to each site. One may even get rid of the lattice completely and look at circles distributed randomly on a piece of paper.

Another important variant, which helps us to go continuously from site to bond percolation, is called *site-bond percolation*. Then the sites of the bond percolation problem are no longer all occupied; only a fraction p of sites is occupied, the rest are empty. Bonds between neighbouring occupied sites are open with probability x, and we look for clusters of occupied sites connected by open bonds. The bond percolation threshold x_c now decreases from unity, if p equals the site percolation threshold, to the normal bond threshold if $p = 1$. See Fig. 21, Section 4.2.

Numerous other modifications of percolation have been invented as models for various processes occurring in nature. The purpose of the present book is to offer an introduction to percolation theory, not a comprehensive review. Thus we will ignore all these complications of connections with next-nearest neighbours, and other processes, and will work with site percolation on various random lattices in d dimensions, except when otherwise stated.

2.2. EXACT SOLUTION IN ONE DIMENSION

Like so many other problems of theoretical physics, the percolation problem can be solved exactly in one dimension, and some aspects of that solution seem to be valid also for higher dimensions.

Let us study site percolation on an infinitely long linear chain, where 'lattice' sites are placed in fixed distances (Fig. 7). Each of these lattice sites is randomly occupied with probability p. A cluster is a group of neighbouring occupied sites containing no empty site in between. A single empty site would split the group into two different clusters. In order that the cluster is separated from the other clusters, the site neighbouring the left end of the cluster must be empty; and the same is true for the right end of the cluster. Thus for the central cluster of Fig. 7 consisting of five occupied sites, we need these five sites occupied and their two neighbours empty.

The probability of each site being occupied is p. Since all sites are occupied randomly, the probability of two arbitrary sites being occupied is p^2, for three being occupied is p^3, and for five being occupied is p^5. (This product property of the probabilities is valid only for statistically independent events, as for random percolation). The probability of one end having an empty neighbour is $(1 - p)$, and again the two ends are statistically independent. Therefore the total probability, that a fixed lattice site is the left end of a five-cluster is $p^5(1 - p)^2$.

How many clusters of size five do we have in our chain, if the total chain length is L, with $L \to \infty$, much larger than the cluster length? Every site has the probability $p^5(1 - p)^2$ of being the left hand end of such a cluster, and there are L such sites (when we ignore the small number of sites on the end of the whole chain for which the situation is different since there no place is left for five occupied and two empty sites). Thus the total number of five-clusters, apart from effects from the chain ends, is $Lp^5(1 - p)^2$. We see that it is practical to talk about the number of clusters per lattice site, which is the total number divided by L and thus $p^5(1 - p)^2$. This normalized cluster

Fig. 7. Example of clusters in a one-dimensional lattice. The central cluster has five sites; the one to its left is a pair; the one to its right is a cluster of size one, that is an isolated occupied site. The empty sites are not shown.

number is thus independent of the lattice size L and equals the probability that a fixed site is the end of a cluster.

For clusters containing s sites, we define n_s as the number of such s-clusters per lattice site. In our one-dimensional case, the above consideration for five-clusters is easily generalized to

$$n_s = p^s(1 - p)^2 \qquad (1)$$

This normalized *cluster number* is crucial for many of our later discussions in two or three dimensions. It equals the probability, in an infinite chain, of an arbitrary site being the left hand end of the cluster. For $p < 1$, the cluster numbers go exponentially to zero if the cluster size s goes to inifinity.

The probability that an arbitrary site is part of an s-cluster, and not only its left end, is larger by a factor s; for now that site can be any of the s cluster sites. Thus that probability is $n_s s$. Many authors prefer to work with the probability $n_s s$, instead of with the cluster number n_s. To avoid confusion with the probability p we will not introduce a special symbol for $n_s s$ and will work with the cluster numbers. This cluster number is also the more natural quantity if one counts all clusters in a lattice of a fixed large size by a computer simulation.

Where is the percolation threshold? For $p = 1$, all sites of the chain are occupied, and the whole chain constitutes one single cluster. For every p smaller than unity, there will be some holes in the chain where a site is not occupied. Thus a chain of length L will have on average $(1 - p)L$ empty sites. For L going to infinity at fixed p, this number is also increasing to infinity. Thus there will be at least one empty site somewhere in the chain, and that means that there is no continuous row of occupied sites, i.e. no one-dimensional cluster, connecting the two ends. In other words, there is no percolating cluster for p below unity. Thus the percolation threshold is unity:

$$p_c = 1 \qquad (2)$$

Therefore it is not possible to observe the region $p > p_c$ in one dimension. Only one side of the phase transition is accessible since at least these authors cannot occupy a site with a probability $p > 1$. Nevertheless, this somewhat unusual phase transition has some similarities with percolation in higher dimensions, and also with certain aggregation processes (Kolb and Herrmann, 1985). Thus we will try to squeeze out some more information from this simple result.

The probability that a site belongs to a cluster of size s is $n_s s$, as discussed above. Every occupied site must belong to one cluster since single occupied sites surrounded by empty neighbours are also clusters of size unity. The probability that an arbitrary site belongs to any cluster is therefore equal to the probability p that it is occupied. Thus

$$\sum_s n_s s = p \qquad (p < p_c) \qquad (3)$$

The sum runs from $s = 1$ to $s = \infty$. This law can also be checked directly from Eq. (1) and the formula for the geometric series:

$$\sum_s p^s (1-p)^2 s = (1-p)^2 \sum_s p \frac{d(p^s)}{dp}$$

$$= (1-p)^2 p \frac{d\left(\sum_s p^s\right)}{dp}$$

$$= (1-p)^2 p \frac{d(p/(1-p))}{dp}$$

$$= p$$

For higher dimensions, Eq. (3) is also valid except that one has to take into account the sites in the infinite cluster separately, if one does not include them in the sum over all cluster sizes. Therefore Eq. (3) is restricted to $p < p_c$; even in one dimension at $p = p_c = 1$ there is only one cluster covering the whole lattice, thus $s = \infty$ and $n_s = 0$, which makes Eq. (3) undefined at $p = 1$. (The above trick to calculate a sum by expressing it as a derivative is also useful in other parts of statistical physics.)

How large on average is the cluster we are hitting if we point randomly to a lattice site which is part of a finite cluster? There is a probability $n_s s$ that an arbitrary site (occupied or not) belongs to an s-cluster and a probability $\Sigma_s n_s s$ that it belongs to any finite cluster. Thus $w_s = n_s s / \Sigma_s n_s s$ is the probability that the cluster to which an arbitrary occupied site belongs contains exactly s sites. The average cluster size S which we are measuring in this process of randomly hitting some cluster site is therefore

$$S = \sum w_s s$$

$$= \sum \frac{n_s s^2}{\sum n_s s} \tag{4}$$

Although we will learn later that different types of averages exist, the term *mean cluster size* is in widespread use for S and will also be used here. (For example, $\Sigma_s n_s s / \Sigma_s n_s$ is the average cluster size if every cluster, and not every site as in Eq. (4), is selected with equal probability.) We have defined S here in such a way that Eq. (4) is also our definition for higher dimensions provided that the infinite cluster is excluded from the sums.

Let us now calculate this mean cluster size explicitly. The denominator is simply p, as Eq. (3) shows. The numerator is

$$(1-p)^2 \sum_s s^2 p^s = (1-p)^2 \left(p \frac{d}{dp}\right)^2 \sum_s p^s$$

where again the sums go from $s = 1$ to infinity, and where the trick from our derivation of Eq. (3) is applied twice in order to calculate sums by using suitable derivatives of easier sums. Thus

$$S = \frac{(1+p)}{(1-p)} \qquad (p < p_c) \qquad (5)$$

The mean cluster size diverges if we approach the percolation threshold. We will obtain similar results later in more than one dimension. This divergence is very plausible, for if there is an infinite cluster present above the percolation threshold, then slightly below the threshold one already has very large (though finite) clusters. Thus a suitable average over these cluster sizes is also getting very large, if one is only slightly below the threshold.

We may define the *correlation function* or pair connectivity $g(r)$ as the probability that a site a distance r apart from an occupied site belongs to the same cluster. For $r = 0$ that probability $g(0)$ equals unity, of course. For $r = 1$ the neighbouring site belongs to the same cluster if and only if it is occupied; this is the case with probability p. For a site at distance r, this site and the $(r-1)$ sites in between this site and the origin at $r = 0$ must be occupied without exception, which happens with probability p^r. Thus

$$g(r) = p^r \qquad (6)$$

for all p and r. For $p < 1$ this correlation function, which is also called a connectivity function, goes to zero exponentially if the distance r goes to infinity:

$$g(r) = \exp\left(\frac{-r}{\xi}\right)$$

where

$$\xi = -\frac{1}{\ln (p)} = \frac{1}{(p_c - p)} \qquad (7)$$

The last equality in Eq. (7) is valid only for p close to $p_c = 1$ and uses the expansion $\ln(1-x) = -x$ for small x. (In many cases we shall be interested only in the asymptotic behaviour near p_c, and then use the equality sign as in Eq. (7). In the literature one also uses the symbol $\simeq$ in such cases.) The quantity ξ is called the *correlation* (or connectivity) *length* and we see that it also diverges at the threshold. We will see later in higher dimensions that the correlation length is proportional to a typical cluster diameter. This relation is quite obvious here. The length of a cluster with s sites is $(s-1)$, not much different from s if s is large. Thus the average length ξ varies as the average cluster size S:

$$S \propto \xi \qquad (p \to p_c) \qquad (8)$$

Unfortunately we will see later that this relation becomes more complicated in higher dimensions. Rather more generally valid is a relation between the sum over all distances r of the correlation function, and the mean cluster size:

$$\sum_r g(r) = S \qquad (9)$$

(The reader who wants to check this and has difficulties should keep in mind that the sum in Eq. (9) not only includes $r = 0, 1, 2, ...$, i.e. the neighbours to the right, but also the neighbours to the left. They cannot be treated through $r = -1, -2$, however, since r is a non-negative distance. Thus one should calculate the sum over the right neighbours and the centre, $r = 0, 1, 2, ...$, multiply it by 2 to take into account the left part of the lattice, and subtract the contribution unity from the centre, which otherwise would be counted twice.)

We see from this exact solution for one dimension, that certain quantities diverge at the percolation threshold, and that the divergence can be described by simple power laws like $1/(p_c - p)$, at least asymptotically close to p_c. The same seems true in higher dimensions where the problems have not been solved exactly.

The quantities S and ξ have counterparts for thermal phase transitions. In fluids near their critical point, critical opalescence is observed in light-scattering experiments, since the *compressibility* (analogous to S) and the correlation length ξ diverge there. The van der Waals equation for fluids in his thesis of 1873 was the first successful theory to describe aspects of such thermal phase transitions.

One may utilize one-dimensional percolation further by calculating the cluster numbers in finite one-dimensional chains. Then one can check the general concepts of finite-size scaling and universality. We leave these problems to Chapter 4.

The one-dimensional case is now solved exactly, whereas for the d-dimensional case only small clusters will be treated exactly in Section 2.3. There is another case with an exactly known solution, the Bethe lattice, with which we deal in Section 2.4.

2.3. SMALL CLUSTERS AND ANIMALS IN d DIMENSIONS

If the one-dimensional solution of Eq. (1) is so simple, why cannot we apply the same principle to higher dimensions and find the exact solution there? To answer that question, let us look again at the square lattice of Fig. 1. First, what is the probability that an arbitrary site is a cluster of size $s = 1$, i.e. an isolated occupied site? For this purpose, the site itself has to be occupied (probability p) while its four neighbour squares have to be empty (probability $(1 - p)$ for each). Again the occupation of these five sites happens independently, and thus the combined probability is the product $n_1 = p(1 - p)^4$. The number of pairs, n_2, can also be calculated easily: Two sites have to be occupied, their six neighbour squares have to be empty, and the pair can be oriented either horizontally or vertically. Thus the average number of pairs per lattice site is $n_2 = 2p^2(1 - p)^6$. Similarly, three sites on a straight line have eight neighbours, and the average number (per lattice site) of such clusters is $2p^3(1 - p)^8$. Generally, the number of clusters of s sites forming a straight line is $2p^s(1 - p)^{2s+2}$ on a square lattice, since each such cluster has $(2s + 2)$ empty neighbour squares.

In three dimensions, on the simple cubic lattice, each straight cluster with s sites has $(4s + 2)$ empty neighbours, and three orientations are possible, leading to an average cluster number (per lattice site) of $3p^s(1 - p)^{4s+2}$.

In a d-dimensional hypercubic lattice, each site has $2d$ neighbours, and for the sites in the interior of an s-cluster forming a straight line, $(2d - 2)$ of these sites have to be empty. Including the two end points, the s-cluster has in this case $2 + (2d - 2)s$ empty neighbours, resulting in a cluster number $dp^s(1 - p)^{(2d-2)s+2}$.

This general d-dimensional result includes the above cluster numbers for $d = 2$ and $d = 3$ as well as the one dimensional result $(d = 1)$ of Eq. (1). Have we thus solved the percolation cluster problem exactly in d dimensions, leaving the evaluations of mean cluster size and correlation length as an exercise analogous to the one-dimensional case?

Unfortunately, our straightforward, exact, simple and complete solution has one slight disadvantage:

IT IS WRONG

The world is not straight. Three sites of a cluster on a square lattice do not necessarily follow a straight line; they can also form a corner, as shown here:

```
        o
      o x o
      o x x o
      o   o
```

The three occupied cluster sites are marked by an x, the seven empty neighbour sites by an o. Four orientations of this corner are possible; thus the average number (per lattice site) of such corners is $4p^3(1 - p)^7$.

Combined with the above result for straight lines we thus get $n_3 = 2p^3(1 - p)^8 + 4p^3(1 - p)^7$ for the average number of triplets on the square lattice. Figure 8 shows the 19 possible configurations for $s = 4$ on the square lattice; it is a nice classroom exercise to find, in a long collaborative effort, all 63 configurations for $s = 5$.

Such an exercise will convince you that the configurations of larger clusters are counted faster and more reliably by a computer. Up to $s = 24$ on the square lattice this has been done by Redelmeier, who kept a PDP 11/70 computer busy for ten months to count the 10^{13} configurations. A short and complete FORTRAN program for this purpose was published by Redner as cited by Mertens (1990). One calls these configurations '*lattice animals*' since they have a certain similarity with multicellular living beings which might enter your nightmares if you counted them too long.

But for our purpose it is not sufficient to count these animals; we have to classify them according to the number of empty neighbours each of them has. For example, of the six triplet configurations, two have eight empty

Fig. 8. List of all cluster configuration ('lattice animals') on the square lattice up to
$s = 4$. For each structure, mirror images and rotated configurations are not shown;
only the total number of such configurations, including the one shown, is indicated
under each structure.

neighbours and the remaining four have seven. This difference entered into
our above result for the average number of triplets. Generally, the number of
empty neighbours of a cluster is called its *perimeter* for which we use the
symbol t here (not to be confused with time, of course). (One should not
identify the perimeter with a cluster surface since t includes internal holes, like
in a Swiss cheese.) Triplets thus have the perimeters $t = 7$ or $t = 8$ on the
square lattice. If the number of lattice animals (cluster configurations) with
size s and perimeter t is denoted by g_{st}, then

$$n_s = \sum_t g_{st} p^s (1 - p)^t \qquad (10)$$

is the average number of s clusters per lattice site. All our above formulae for
cluster numbers are special cases of this general formula, which is valid for
every lattice. For example, the triplets ($s = 3$) on the square lattice have
perimeter $t = 8$ ($g_{st} = 2$ configurations) or $t = 7$ ($g_{st} = 4$ configurations), and
the total cluster number is the sum of these contributions.

The difficulty with Eq. (10) is that it involves a sum over all possible perimeters t, and thus each possible configuration has to be found and carefully
analysed to find the g_{st}. Tables of such animal numbers, often in the form of
so-called perimeter polynomials

$$D_s(q) = \frac{n_s}{p^s} = \sum_t g_{st} q^t \qquad \text{where } q = (1 - p)$$

have been published mainly by the King's College group (see Further
Reading). There seems to be no exact solution for general t and s available
at present, and that is why the percolation cluster problem has not yet been
solved exactly.

Nevertheless, some asymptotic results are known for very large animals.
They are listed here without proof since according to our present knowledge
these animals, even if domesticated by exact solutions, do not help us in an
exact solution for percolation clusters at the threshold. The perimeter t, averaged over all animals with a given size s, seems to be proportional to s for
$s \to \infty$. Thus it is appropriate to classify different animals of the same large
size s by the ratio $a = t/s$. If a is smaller than $(1 - p_c)/p_c$ on any lattice in

more than one dimension, then g_{st} varies as

$$\left[\frac{(a+1)^{a+1}}{a^a}\right]^s \tag{11}$$

for large animals, apart from factors varying less strongly with s (B. Souillard, unpublished; see the reviews of Stauffer (1979) and Essam (1980) cited in Chapter 1). Therefore also the total number $g_s = \Sigma_t\, g_{st}$ of animals, irrespective of their perimeter, increases exponentially with animal size s, apart from less strongly varying prefactors:

$$g_s \propto s^{-\theta}\, \text{const}^s \tag{12}$$

In two dimensions $\theta = 1$, whereas $\theta = 3/2$ in three (Parisi and Sourlas, 1981); $\theta = 5/2$ for d above 8, as in the Bethe lattice. Finally, the average radius or diameter of big animals increases as the square root of the animal mass s for large s in three dimensions, in contrast to the impression you may have got in the zoo. (No exact solution for two-dimensional animal radii is known thus far, one of the rare cases where $d = 3$ is better known than $d = 2$.)

Inspection of Eq. (10) tells us that averages over percolation clusters of one fixed size correspond, in the limit $p \to 0$, to average over lattice animals. For then the factor $(1 - p)^t$, by which percolation clusters with different perimeters t are distinguished from animals, approaches unity and thus can be omitted. Thus for very small p, the average squared radius of percolation clusters, discussed later in Section 3.2, also varies as s for large s in three dimensions. The numbers of such large percolation clusters are very small, of course, as follows from Eq. (12):

$$n_s(p \to 0) \propto s^{-\theta} p^s\, \text{const}^s \tag{13}$$

which goes to zero rapidly for increasing s if $p \ll 1/\text{const}$. Thus the percolation clusters to which our animal limits apply are very rare.

2.4. EXACT SOLUTION FOR THE BETHE LATTICE

Besides the one-dimensional case, another case can also be solved exactly which in some sense corresponds to infinite dimensionality: the *Bethe lattice* (or *Cayley tree*) of Fig. 9. The Bethe approximation is a method used to treat magnetism and works exactly on Cayley trees; that is why physicists call these structures 'Bethe lattices'.

What has the Bethe lattice to do with infinite dimensionality d? For $d = 2$, the area of a circle with radius r is πr^2 whereas its circumference is $2\pi r$. The surface of a three-dimensional sphere of radius r is $4\pi r^2$ whereas its volume varies as r^3. In d dimensions, the volume of a 'sphere' is proportional to r^d, its surface to r^{d-1}. Thus

$$\text{surface} \propto \text{volume}^{(1 - 1/d)} \tag{14}$$

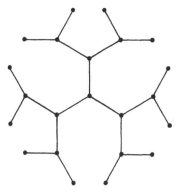

Fig. 9. A small Cayley tree or Bethe lattice, where each site except the many surface sites has $z = 3$ neighbours. Percolation theory started with the exact solution on this somewhat artificial structure.

in d dimensions. We see that in the limit $d \to \infty$ the surface becomes proportional to the volume; this is also true if we look at squares, cubes, 'hypercubes' etc.

Another argument concerns the probability of finding loops in high dimensions. We can demonstrate this by looking at clusters of four sites. For very large d, the number of ways to embed a chain of four sites on a hyper-cubic lattice is proportional to $(2d - 1)^3$ (each of the last three sites can be placed along another axis). However, the number of ways to have a 'loop', in which the four sites sit on the same plane and form a square, is proportional to $d(d - 1)$. Therefore, loops become relatively unimportant when $d \to \infty$, and the results are the same as on 'trees'.

For the Bethe lattice in Fig. 9, one starts with a central point ('origin'), having z bonds, with $z = 3$ in the example of Fig. 9. Each bond ends in another site from which again z bonds emanate; one of these z bonds is the connection with the origin, the other $(z - 1)$ bonds lead to new sites. This branching process is continued again and again. Thus if we have reached one site in the interior of a Bethe lattice, then we can go on in $(z - 1)$ other directions in addition to the direction from which we came. Only at the surface of the lattice, where the branching has stopped, is only one bond connecting the surface site to the interior ('dead end road'). There are no closed loops in this structure, which means that we always reach new sites if we never go back.

We see from Fig. 9 that the number of sites increases exponentially with the distance from the origin, whereas in any d-dimensional structure it would increase with (distance)^d. In our example with $z = 3$, the origin is surrounded by a shell of three sites ('first generation'), in the second shell we have six sites, followed by a third generation of twelve sites, etc. Thus a sphere of r generations contains $1 + 3(1 + 2 + 4 + \cdots + 2^{(r-1)}) = 3 \times 2^r - 2$ sites, of which the last generation of $3 \times 2^{(r-1)}$ sites are surface sites. Thus for large r half of the sites are surface sites, the other half are in the interior of the sphere. With z instead of three neighbours for each site, the fraction of surface sites approaches $(z - 2)/(z - 1)$, as an analogous calculation shows. Thus the ratio of surface to volume approaches a finite limit. Equation (14) shows that this special behaviour occurs for $1/d = 0$ only, that is for infinite dimensions. We thus see that the Bethe lattice is something very peculiar. When we now talk about percolation in the Bethe lattice, we therefore always have in mind the behaviour in the interior of the Bethe lattice, and not the effects due to the surface, which are also important.

Let us now find the percolation threshold in the Bethe lattice. We start at the origin and check if there is a chance of finding an infinite path of occupied neighbours, starting from that origin. If we go on such a path in the outward direction, we find $(z - 1)$ new bonds emanating from every new site, apart from the direction from which we came. Each of these $(z - 1)$ bonds leads to one new neighbour, which is occupied with probability p. Thus on average we have $(z - 1)p$ new occupied neighbours to which we can continue our path. If this number $(z - 1)p$ is smaller than unity, the average number of different paths leading to infinity decreases at each generation by this factor < 1. Thus, even if all z neighbours of an occupied origin happen to be occupied, giving us z different chances to find a way out, and even if z is very large, the probability of finding a continuous path of occupied neighbours goes to zero exponentially with path length if $p < 1/(z - 1)$. Therefore we have derived

$$p_c = \frac{1}{z - 1} \tag{15}$$

for the Bethe lattice with z neighbours for every site. (The above argument is also valid if each bond between neighbour sites is open or blocked randomly; thus Eq. (15) is valid for both bond percolation and site percolation.)

Even if p is larger than the percolation threshold $1/(z - 1)$, the origin does not always have a connection to infinity. For example, if it is occupied and its z neighbours are empty, then it does not belong to the infinite network. We define the percolation probability P as the probability that the origin or any other arbitrarily selected site belongs to the infinite cluster. Clearly this probability is zero for p below the percolation threshold p_c, and we want to calculate it therefore only for $p > 1/(z - 1)$. As we saw in Section 1.3, this quantity P also makes sense for general lattices, not only for Bethe lattices. (In one dimension our p is never $> p_c$ and thus we did not introduce it there.) To distinguish the two probabilities P (probability of an arbitrary site belonging to the infinite network) and p (probability of an arbitrary site being occupied) we may also call P the *strength* of the infinite network and p the concentration. 'Strength' here means only the relative amount and should not be confused with the elastic property.

Figure 10 shows the immediate surroundings of the origin and defines what we mean by branch, neighbour, and *subbranch*. We want to calculate

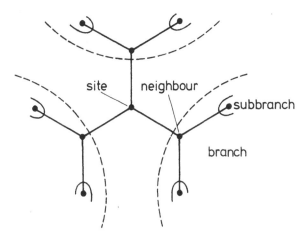

Fig. 10. The surroundings of the origin of a Bethe lattice. This figure defines what we mean by neighbour, branch and subbranch in our derivation of the exact solution.

the strength of the infinite network, that is the probability P that the origin (or any other site) is connected to infinity by occupied sites. We call Q the probability that an arbitrary site is not connected to infinity through one fixed branch originating from this site. Taking $z = 3$ for simplicity, as we have done in Fig. 10, we now calculate Q from the rule that probabilities for statistically independent events are simply multiplied by each other. The probability that the two subbranches which start at the neighbour are not both leading to infinity is Q^2. (A subbranch is connected to infinity with the same probability as a branch since all sites are equivalent in the interior of the Bethe lattice.)

Thus pQ^2 is the probability that this neighbour is occupied but not connected to infinity by any of its two subbranches. This neighbour is empty with probability $(1 - p)$, in which case even well connected subbranches do not help it. Thus $Q = 1 - p + pQ^2$ is the probability that this fixed branch does not lead to infinity, either because the connection is already broken at the first neighbour, or because later something is missing in a subbranch.

This quadratic equation for Q has two solutions $Q = 1$ and $Q = (1 - p)/p$. The probability $(p - P)$ that the origin is occupied but not connected to infinity through any of its three branches is pQ^3. Thus $P = p(1 - Q^3)$, which gives zero for the solution $Q = 1$ (apparently belonging to $p < p_c = 1/2$), and gives

$$\frac{P}{p} = 1 - \left[\frac{1-p}{p}\right]^3 \tag{16}$$

for the other solution, which corresponds to $p > p_c = 1/2$.

Figure 11 displays this result, which goes back to Flory (1941), for it is in polymer chemistry that the first percolation theory was developed by studying bond percolation on this Bethe lattice. (As we saw above, for this

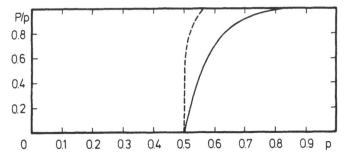

Fig. 11.   Order parameter P (strength of the infinite network) versus concentration p, in the Bethe lattice with $z = 3$. From equation (16) shown as a solid line and compared with computer simulations of the triangular lattice (dashed line). In both cases the threshold is at $p_c = 1/2$.

special case the difference between bond and site percolation is not very important.) P is then identified with the fraction of atoms which belong to the infinite network. For example, if one prepares a pudding it is first a fluid (no elasticity, finite viscosity, finite macromolecules, finite clusters, $p < p_c$). After some time it is a jelly with a finite elasticity, and no longer fluid. This process is called *gelation* and is also observed if an egg is boiled for breakfast, if milk becomes cheese, if rubber is vulcanized etc. Flory identified this polymerization process with percolation and solved the percolation problem by using the Bethe lattice, which approximates reality by not allowing any closed loops. Indeed, certain Scottish cows have learned since then to follow Eq. (16) quite closely when their milk is transformed to cheese. For more gelation details we refer to the review of Kolb and Axelos (1990) listed after Chapter 1.

Just as in one dimension, we can also calculate the mean cluster size S for the Bethe lattice. This is the average number of sites of the cluster to which the origin belongs. Again we take $z = 3$ for simplicity. Let T be the mean cluster size for one branch, that is the average number of sites to which the origin is connected and which belongs to one branch. Again, subbranches have the same mean cluster size T as the branch itself. If the neighbour is empty (probability $(1 - p)$), the cluster size for this branch is zero. If the neighbour is occupied (probability p), it contributes its own mass (unity) to the cluster, and adds the mass T for each of its two subbranches. Thus,

$$T = (1 - p)0 + p(1 + 2T)$$

with the solution $T = p/(1 - 2p)$ for p below the threshold 1/2. The total cluster size is zero if the origin is empty and $(1 + 3T)$ if the origin is occupied; therefore the mean size is

$$S = 1 + 3T = \frac{1 + p}{1 - 2p} \tag{17}$$

We have thus derived exact formulae for the mean cluster size S below the percolation threshold (Eq. (17)) and the strength P of the infinite cluster for concentrations p above the percolation threshold (Eq. (16)).

If there is an infinite cluster above p_c it is plausible that slightly below p_c the mean cluster size is very large, and indeed the denominator of Eq. (17) vanishes for $p = 1/2 = p_c$, giving

$$S \propto \frac{1}{p_c - p} \qquad (18a)$$

if p approaches p_c from below. On the other hand, if there is no infinite network below p_c it is very possible that slightly above p_c one has only a very weak infinite network, that is P is very small. Indeed, Eq. (16) tells us that $P = 0$ at $p = 1/2$, and

$$P \propto (p - p_c) \qquad (18b)$$

if p approaches p_c from above. (To obtain the proportionality factor in Eq. (18b) you should make a Taylor expansion of Eq. (16) in powers of $p - \frac{1}{2}$. We recommend that you do this as an exercise; see exercise 2.5, p. 174. Note that this Taylor expansion also gives corrections of order $(p - p_c)^2$, which can be used to estimate the range of validity of Eq. (18b)).

Equation (18) is an example of critical phenomena: Quantities of interest go to zero or infinity by simple *power laws*. We will discuss similar power laws later when we go to regular d-dimensional lattices instead of the Bethe lattice, only then the power laws are not so simple, with P for example vanishing as $(p - p_c)^{5/36}$ in two dimensions. The power laws are so particularly simple in the Bethe lattice because it is exactly solved (you may also see it the other way round); a simple formula like Eq. (16) can hardly give a critical exponent $5/36$.

Critical phenomena also occur for thermal phase transitions; the Bethe lattice approximation for percolation theory then is somewhat analogous to the molecular field approximation for magnetism, or the van der Waals equation for fluids. (More precisely, one should use the better Bethe approximation for magnetism, which is exact on the Bethe lattice but has the same critical exponents as the molecular field approximation.) In all three cases, rather simple formulae for the order parameter can be derived exactly, leading to a simple power law very near to the critical point. For example, the difference between liquid and vapour density, or the spontaneous magnetization, vanish in both approximations as $(T_c - T)^{1/2}$. Thus the analogy between thermal critical phenomena and percolation is not complete since the critical exponent for the order parameter is $1/2$ for thermal phase transitions and unity for percolation, in these approximations.

In both the thermal phase transitions and percolation, the order parameter goes to zero continuously if one approaches the critical point. Such phase transitions are called continuous phase transitions or second-order phase transitions. If instead the order parameter jumps to zero, one has a first-order phase transition. Such transitions can occur in more complicated situations, like 'bootstrap percolation' (as reviewed by Adler, 1991), where on a square lattice a site remains occupied only if three or four of its neighbours are still occupied.

The above derivations make clear why it is important to have a Bethe lattice with no loops as in the square lattice but only branches as in a tree. If there had been any connections between the different branches or sub-branches, except for the site or neighbour where the (sub-)branches separate, then we could not have calculated the probabilities for different branches together as products of probabilities for each branch separately. QQQ would no longer have been the probability that none of the three branches is connected to infinity, had the three branches not been statistically independent.

Another reason why Bethe lattices are easier to solve than square lattices becomes clear if one calculates $n_s(p)$, the average number (per site) of clusters containing s sites each. As in one dimension and in contrast to two or three dimensions, the size s of a cluster is uniquely related to its perimeter t, that is to the number of empty neighbours of occupied cluster sites. A single isolated site has three empty neighbours in Fig. 9, a pair has four. In the general case of z neighbours per site, the isolated bachelor is surrounded by z perimeter sites whereas the married couple has $(2z - 2)$ empty neighbours. Each child added to that cluster increases t by $(z - 2)$. Thus $t = (z - 2)s + 2$ is the total perimeter of s-clusters in the Bethe lattice, just as in the square or simple cubic lattices for straight-line clusters. We see that for large s the perimeter is proportional to s, as is also the case in more realistic lattices. Moreover, the asymptotic ratio t/s equals $(1 - p_c)/p_c$ since this ratio is $(z - 2)$ and since $p_c = 1/(z - 1)$. We will see later (Section 3.1) that this relation is valid at the percolation threshold for all lattices, not only for the Bethe lattice.

Now we apply our general result given by Eq. (10) for the cluster numbers, inserting $t = (z - 2)s + 2$:

$$n_s = g_s p^s (1 - p)^{2 + (z-2)s}$$

(Since for each s only one value for t is possible we do not have to sum over t.) Since we are lazy, we set $z = 3$ for simplicity (similar results hold generally), avoid the calculation of the number g_s of different configurations for s-clusters on the Bethe lattice, and instead look at the ratio

$$\frac{n_s(p)}{n_s(p_c)} = \left[\frac{(1 - p)}{(1 - p_c)}\right]^2 \left[\left(\frac{p}{p_c}\right)\frac{(1 - p)}{(1 - p_c)}\right]^s$$

$$= \left[\frac{(1 - p)}{(1 - p_c)}\right]^2 [1 - a(p - p_c)^2]^s$$

$$\propto \exp(-cs) \qquad\qquad (19)$$

Here

$$a = 4 \qquad c = -\ln[1 - a(p - p_c)^2] \propto (p - p_c)^2$$

We see that a very simple exponential decay is obtained in Eq. (19) for this ratio of cluster numbers. Later we will see that this simplicity is a peculiarity of the Bethe lattice. In two or three dimensions a decay law as $\exp(-cs)$ is valid only for large clusters, and only for $p < p_c$. We call this decay as $\exp(-cs)$ 'animal-like' since for very small p, that is in the animal limit of Section 2.3, the cluster numbers decay as $(p \times \text{const})^s$, Eq. (13). The Bethe lattice clusters are always animal-like, whereas in two or three dimensions only very large percolation clusters below p_c behave like animals.

We now want to find the asymptotic behaviour of the cluster numbers at the threshold, $n_s(p_c)$. We have seen generally, not only for Bethe lattices, in Eq. (4) that

$$S \propto \sum_s s^2 n_s$$

since the denominator remains finite at the threshold. Thus for $p = p_c$ this sum (also called the second moment of the cluster size distribution) is infinite, whereas for any other p it remains finite. If $n_s(p_c)$ decayed exponentially with s, then the mean cluster size S would remain finite at $p = p_c$. Thus a power law decay is more plausible and defines the Fisher exponent τ (Fisher droplet model; Fisher, 1967) through

$$n_s(p_c) \propto s^{-\tau} \tag{20}$$

for large s. Again this law is valid rather more generally, not only in the Bethe lattice.

Let us now evaluate S and then calculate τ by comparing the result with Eq. (18). The calculation which follows now uses tricks which occur again and again in the scaling theory of percolation clusters. We assume p to be only slightly smaller than p_c:

$$\begin{aligned}
S &\propto \sum s^2 n_s \\
&\propto \sum s^{2-\tau} \exp(-cs) \\
&\propto \int s^{2-\tau} \exp(-cs)\, ds \\
&= c^{\tau-3} \int z^{2-\tau} \exp(-z)\, dz \\
&\propto c^{\tau-3} \\
&\propto (p_c - p)^{2\tau-6}
\end{aligned} \tag{21}$$

where we used the above result that c vanishes quadratically in $p - p_c$. Equation (18) shows that S must diverge with an exponent -1:

$$2\tau - 6 = -1$$

Thus the Fisher exponent is

$$\tau = 5/2 \tag{22}$$

in the Bethe lattice. We thus can rewrite Eq. (19) for large s:

$$n_s(p) \propto s^{-5/2} \exp(-cs) \qquad c \propto (p - p_c)^2 \tag{23}$$

where the first proportionality holds for all p and large s, the second one only for p near p_c.

Of course one can also derive Eq. (23) differently (see for example Essam's 1980 review). It is even possible to calculate the cluster numbers at the threshold exactly by calculating the number g_{st} of 'animals' of the Bethe lattice. The resulting expression involving binomial coefficients is, however, less useful for our later studies of two and three dimensions than the above derivation.

Having derived the Fisher exponent $\tau = 5/2$ from the critical behaviour of the mean cluster size S, we can check whether it also gives the correct critical behaviour (Eq. (18)) of P, the strength of the infinite network. For this purpose we use a general equation, valid for all site percolation problems, not only in the Bethe lattice:

$$P + \sum_s n_s s = p \tag{24}$$

where the sum runs over all finite cluster sizes s and excludes the infinite cluster. This equation simply states that all occupied sites (probability p) either belong to the infinite cluster (probability P) or to one of the finite clusters (probability $\sum n_s s$); it generalizes our one-dimensional result, Eq. (3), to $p > p_c$ where an infinite network has to be taken into account. (Note again that isolated sites are regarded as clusters of size $s = 1$.) You may now check Eq. (18) yourself, finding help in Eqs. (28) and (29) using $\sigma = 1/2$ and $\tau = 5/2$.

(Let us mention in passing that besides the Bethe lattice approximation there also exists the effective medium approximation, in particular for percolation properties other than cluster numbers. The critical exponents of this approximation agree neither with those of three-dimensional lattices nor with those of the Bethe lattice, but the behaviour predicted somewhat away from the percolation threshold is quite realistic. For details we refer to Kirkpatrick's (1973) review article mentioned after Chapter 1.)

In summary, the Bethe lattice solution, Eqs. (18) and (23) as well as the one-dimensional solution, Eq. (1), show that cluster numbers follow rather simple laws, and that exponential decay is quite common. We will utilize these results in the next section to make plausible the scaling law for cluster numbers in general, not just for one- and infinite-dimensionality.

2.5. TOWARDS A SCALING SOLUTION FOR CLUSTER NUMBERS

In this section we try to invent a simple formula which contains the previously discussed one-dimensional and Bethe lattice solutions as special cases. While that formula still is not the final one to be discussed in Section 2.6, it already has many of its properties and in particular already yields the desired scaling laws.

Both Eq. (1) for one dimension and Eq. (23) for 'infinite-dimensionality' (Bethe lattice) are dominated for large cluster sizes s by a simple exponential decay law, $\log(n_s) \propto -s$. Thus we may first postulate for d-dimensional percolation:

$$n_s \propto \exp(-cs)$$

where the factor of proportionality and the parameter c depend on the concentration p. Indeed, if we use the exact result for $s = 1, 2, 3$ in the square lattice mentioned at the beginning of Section 2.3, we find for $p = 0 \cdot 1$ the cluster numbers n_s to be $0 \cdot 06561$, $0 \cdot 01063$, and $0 \cdot 00277$ for $s = 1, 2$, and 3, indicating at least a very rapid decay. However, this nice exponential is not consistent with Eq. (23), where for the Bethe lattice we also found, in contrast to the one-dimensional case, a power law factor $s^{-\tau}$ by which the exponential is multiplied. Thus instead we postulate more generally:

$$n_s \propto s^{-\tau} \exp(-cs) \qquad (25)$$

This law is supposed to be valid for large s only, as was Eq. (23). Again, the proportionality factor and the parameter c depend on p, whereas we assume the exponent τ to be a constant, though not necessarily equal to $5/2$ as in Eq. (23). Moreover, near the percolation threshold we no longer require c to vanish as $(p - p_c)^2$ but instead allow a more general power law:

$$c \propto |p - p_c|^{1/\sigma} \qquad (p \to p_c) \qquad (26)$$

Here σ is another free exponent, not necessarily equal to $1/2$ as in the Bethe lattice solution given by Eq. (23).

Note that the exponential factor in Eq. (25) acts as an effective *cutoff* on the cluster sizes: only clusters with $s < s_\xi = 1/c \propto |p - p_c|^{-1/\sigma}$ contribute significantly to cluster averages. For these clusters, n_s is effectively equal to $n_s(p_c) \propto s^{-\tau}$. Clusters with $s \gg s_\xi$ are exponentially rare, and their properties are no longer dominated by the behaviour at p_c. The size s_ξ can thus also be identified as representing a *crossover* from the behaviour of 'critical' clusters to that of 'non-critical' ones.

Obviously, Eqs. (25) and (26) are a generalization of our results for the Bethe lattice; does it also contain the one-dimensional result given in Eq. (1) as a special case? Using $p_c = 1$ and $p = \exp(\ln p) = \exp(p - 1) = \exp(p - p_c)$ for $p \to p_c$, we rewrite Eq. (1) as

$$n_s(p) = (p_c - p)^2 \exp(-(p_c - p)s)$$

Obviously, this result is not a special case of Eq. (25) since instead of a power of s we have a power of $p - p_c$ in front of the exponential. We shift the resolution of this discrepancy to the next section by beginning in a more general way. Now we have to choose between a generalization of the one-dimensional result and our generalization Eq. (25) of the Bethe lattice solution. We prefer the Bethe lattice as being more realistic than a one-dimensional chain in that it has at least a percolative phase transition: p_c is smaller than unity, in contrast to one dimension, and thus both sides of the threshold can be reached. Thus, for the time being we will work with Eq. (25).

Equation (25) was assumed to be valid for large s only; perhaps we can get rid of some of the deviations for smaller clusters by investigating the ratio $n_s(p)/n_s(p_c)$, which we call $v_s(p)$ instead of n_s. Then Eq. (25) reads

$$v_s \propto \exp(-cs) \qquad (27)$$

where now the exponent τ has cancelled out; it is only implicitly contained in Eq. (27) since, from Eq. (25),

$$n_s(p_c) \propto s^{-\tau}$$

as in Eq. (20). Equation (27) is so simple that it is perhaps worth being trusted by the reader; at least even though not entirely correct it allows us to make many calculations quite easily. (You might violate the Official Secrets Act if you now conclude and say loudly that this is what theoretical physicists do; make calculations if they are easy irrespective of whether the assumptions are correct or wrong.)

First, let us calculate the fraction P of sites belonging to the infinite network. A site is either empty or occupied, and if it is occupied it belongs either to a finite cluster (including isolated sites which are treated as clusters of size $s = 1$) or to the infinite network. If we simply set $s = \infty$ in Eq. (25) we get zero, but that has to be expected. In an infinitely large lattice, which contains at most one infinite network, the number of infinite networks per lattice site is indeed zero. The fraction of lattice sites in the infinite network is calculated by subtracting from the occupied sites those belonging to finite clusters, that is those described by Eq. (24). Right at the critical point $p = p_c$ we have $P = 0$ and thus $\sum_s n_s s = p_c$. To have this sum converge we need $\tau > 2$. Then we rewrite Eq. (24) as

$$P = \sum_s [n_s(p_c) - n_s(p)]s + O(p - p_c)$$

$$\propto \sum_s s^{1-\tau}[1 - \exp(-cs)] \qquad (28)$$

If p is close to p_c, the factor c in the exponent will be quite small, and only large s values of the order of $1/c$ will give the main contribution to the sum. We may therefore replace the sum by an integral if we are interested only in the leading behaviour near the threshold:

$$P \propto \int s^{1-\tau}[1 - \exp(-cs)] \, ds$$

Integration by parts tells us that $\int f'g \, ds = -\int fg' \, ds + (fg)$; here we take

$$f(s) = s^{2-\tau} \qquad g(s) = 1 - \exp(-cs)$$

and get with $z = cs$:

$$P \propto c \int s^{2-\tau} \exp(-cs) \, ds = c^{\tau-2} \int z^{2-\tau} \exp(-z) \, dz$$

(In this definite integral from zero to infinity, the term fg of our integration by parts vanishes since $\tau < 3$). The integral over z is known as the gamma function $\Gamma(3 - \tau)$ and is available in tabular form. For some applications the reader might need the general rule $\Gamma(x + 1) = x\Gamma(x)$ and $\Gamma(1) = 1$, but here we do not need to know anything about gamma functions since the whole integral is simply a numerical factor (note that τ was assumed to be constant):

$$P \propto c^{\tau-2} \propto (p - p_c)^{(\tau-2)/\sigma} = (p - p_c)^{\beta}$$

with the critical exponent

$$\beta = \frac{\tau - 2}{\sigma} \tag{29}$$

Thus we have found that the much simpler result of the Bethe lattice solution given by Eq. (18b) is not valid generally. Instead a critical exponent β is introduced which describes how the strength of the infinite network goes to zero at the percolation threshold.

Secondly, let us calculate how the mean cluster size S diverges at the threshold. As for Eq. (21), we have $S \propto \Sigma s^2 n_s$ since the denominator in Eq. (4) remains finite (Eq. (24)). The same techniques can be applied, but now we may even avoid the integration by parts:

$$S = \sum_s \frac{s^2 n_s}{p_c}$$

$$\propto \int s^2 n_s \, ds$$

$$\propto \int s^{2-\tau} \exp(-cs) \, ds$$

$$\propto c^{\tau-3} \int z^{2-\tau} \exp(-z) \, dz$$

(The careful reader will notice that our calculation neglects the influence of the single infinite cluster. All sums over all cluster sizes from now on are understood to exclude the infinite cluster, if one is present. For one dimension in Eq. (4) we did not need that warning since there one cannot have $p > p_c$.) Again the integral over $z = cs$, which equals $\Gamma(3 - \tau)$, is less interesting than the exponent γ for the divergence of S:

$$S \propto c^{\tau-3}$$

$$\propto |p - p_c|^{(\tau-3)/\sigma}$$

$$= |p - p_c|^{-\gamma}$$

Thus

$$\gamma = \frac{3 - \tau}{\sigma} \tag{30}$$

gives the critical exponent for the mean cluster size S. In order that both β and γ are positive we need $2 < \tau < 3$. For thermal critical phenomena an analogous quantity is the susceptibility of magnets or the compressibility of fluids;

both diverge at the critical point with an exponent γ. The numerical value of γ can, of course, be different for different phase transitions. In general the exponent γ is not a simple number like unity for percolation or for thermal critical phenomena. This is in contrast to Eq. (18a) for the Bethe lattice and the analogous Curie–Weiss law for the mean-field approximation of the susceptibility.

Does this mean that we have to introduce a new exponent about which we know nothing for every new variable? Obviously this is not the case; for all other quantities derived in this way from the cluster numbers n_s have σ and τ as free parameters in the exponents, but nothing else. Thus if we know these two exponents, we know all others. More explicitly, let us calculate the sum

$$M_k = \sum_s s^k n_s \qquad (31a)$$

which experts also call the kth *moment* of the cluster size distribution n_s (if k is an integer). The mean cluster size corresponds to $k = 2$, the strength of the infinite cluster to $k = 1$, and we now allow k to be an arbitrary number $> (\tau - 1)$.

$$M_k \propto \sum_s s^{k-\tau} \exp(-cs)$$

$$\propto \int s^{k-\tau} \exp(-cs)\, ds$$

$$= c^{\tau-1-k} \int z^{k-\tau} \exp(-z)\, dz$$

Thus, apart from a gamma function incorporated into the proportionality factor, we have

$$M_k \propto c^{\tau-1-k} \propto |p - p_c|^{(\tau-1-k)/\sigma} \qquad (31b)$$

Thus the exponent $(k + 1 - \tau)/\sigma$ is again expressed through σ and τ and therefore is not independent of σ and τ. In fact, instead of σ and τ we may also regard β and γ as the fundamental exponents and calculate from them

$$\sigma = 1/(\beta + \gamma) \qquad \tau = 2 + \beta/(\beta + \gamma)$$

As the reader can easily check, this is the solution of Eqs. (29) and (30). Thus the critical exponent for the kth moment is $\beta - (\beta + \gamma)(k - 1)$ according to Eq. (31b). Setting $k = 1$ we recover the exponent β for the strength of the infinite network, whereas for $k = 2$ we get $-\gamma$ for the mean cluster size, as it should be.

Some caution is necessary if a sum is not diverging, that is if $k < \tau - 1$. We have this problem already in the evaluation of the first moment; to get the strength P of the infinite network we had to subtract from the sum its value at $p = p_c$, and then replace the sum by an integral. A simple and more general way is to calculate the first derivative (or second, third, ... derivative, if needed) of the desired sum with respect to c or p. If that derivative diverges, one can replace the sum by an integral, evaluate the result with Eq. (31b), and then go back to the original sum. For example, for the first moment ($k = 1$), Eq. (31b) cannot be applied directly since the sum does not diverge. Instead, we calculate

$$-\frac{\mathrm{d}M_1}{\mathrm{d}c} = \sum s^2 n_s = M_2 \propto c^{\tau-3}$$

$$M_1 \propto \mathrm{const} + c^{\tau-2}$$

in agreement with what we derived immediately before Eq. (29).

For the zeroth moment $M_0 = \sum n_s$ (the total number of clusters), we take the second derivative:

$$\frac{\mathrm{d}^2 M_0}{\mathrm{d}c^2} = M_2 \propto c^{\tau-3}$$

Thus,

$$M_0 = \mathrm{const}_1 + \mathrm{const}_2 c + \mathrm{const}_3 c^{\tau-1}$$

The '*singular*' or non-analytic part of the total number M_0 of clusters thus varies as

$$M_{0_{\mathrm{sing}}} \propto c^{\tau-1}$$

$$\propto |p - p_c|^{(\tau-1)/\sigma}$$

$$\propto |p - p_c|^{(2-\alpha)}$$

with

$$2 - \alpha = \frac{\tau-1}{\sigma} = 2\beta + \gamma \tag{32}$$

(Are you doubtful about whether or not you are allowed to evaluate a sum by replacing it directly with an integral? One way to clarify that question is simply to try it. If it does not work you will notice that fact by realizing that the final integral over z (which should be a constant incorporated into the factor of proportionality) does not exist because it diverges at the lower boundary. In that case you should look at suitable derivatives before replacing the sum by an integral.) If you question the accuracy of replacing sums by integrals you might look at the exact example

$$\sum_{s=1}^{\infty} s \exp(-cs) = \frac{\exp(-c)}{[1 - \exp(-c)]^2}$$

as used already after Eq. (3). Approximating the sum by an integral as above we get $1/c^2$. The exact result is more complicated, but for very small c (and this is what we are looking at in a theory of critical exponents), the exact formula can be expanded into $(1 - 5c/12 + \cdots)/c^2$. Thus, the leading term is calculated correctly by our approximation.)

Unfortunately, there is something wrong with our assumption. Not only does it fail to include the one-dimensional solution as a special case. It also does not work properly for the strength of the infinite network. Nowhere in our above derivation of the exponent did we actually assume $p > p_c$; thus, our formulae would predict an infinite network both above and below the percolation threshold, vanishing only at p_c. Clearly this is wrong, and will be corrected in the following section. To do so we will have to avoid any assumption leading to a maximum of $n_s(p)$ at $p = p_c$ for a fixed large s, when considering the ratio v_s in Eq. (27). This maximum must be located below the percolation threshold, for if it is at p_c then $n_s(p_c) - n_s(p)$ is positive both above and below p_c, giving a non-zero strength P of the infinite network (see derivation after Eq. (27)) even below p_c.

There is another property of our assumptions in Eqs. (25) and (26) which should make us suspicious. In Section 2.3 we learned that for a fixed cluster size s the number $n_s(p)$ of such clusters is a finite polynomial in p. Neither $n_s(p)$ nor any of its p-derivatives is allowed to diverge at p_c. But from Eq. (26) we find divergences in p-derivatives if $1/\sigma$ is not an integer. For example, if σ is close to $0 \cdot 4$ as in two-dimensional percolation, then $c \propto |p - p_c|^{2 \cdot 5}$, and the third derivative of c and thus of n_s with respect to p diverges roughly as $1/|p - p_c|^{1/2}$. A more reliable assumption therefore has to avoid the expression $z = cs \propto |p - p_c|^{1/\sigma}s$ as an argument of the exponential. Instead we may try $z \propto (p - p_c)s^\sigma$ and replace Eq. (27) by $v_s \propto \exp(-z)$. This assumption is known as the Fisher droplet model and is numerically quite good above the percolation threshold. The whole analysis above can be repeated easily with this droplet model formula; basically only the arguments of the gamma functions are changed, which enter the proportionality factors only. Historically this approach was one of the first scaling theories of thermal critical phenomena and also helped in the application to percolation. As desired, $n_s(p)$ now is perfectly smooth at p_c. But still the Fisher model can hardly be correct below p_c since now $n_s(p)$ goes to infinity, instead of zero, for $s \to \infty$.

Nevertheless, our approximation shows the essentials of modern phase transition theory. Everything depends on only two critical exponents. It does not matter whether we call them σ and τ, or β and γ; we only have to keep in mind that from two exponents we can derive the others. For example, from β and γ we can derive the 'specific heat' exponent α via $(2 - \alpha) = (2\beta + \gamma)$. These relationships, known as *scaling laws*, have been used since the 1960s for thermal phase transitions and in the 1970s were extended to percolation theory. By going through the above formalism the reader will have a better feeling for the more general derivations which follow in the next section.

2.6. SCALING ASSUMPTION FOR CLUSTER NUMBERS

If you have read this far through the book it is presumably too late for you to return it and get a refund. Thus now we can tell you the truth: we are unable to offer you the exact solution for the cluster numbers. Instead you are merely offered a further generalization of the above assumption, involving a scaling function so general that everything we discussed so far is contained in it as a special case. No deviations from this scaling assumption have been found (yet) for usual percolation in two and three dimensions.

What have the Fisher model formula, $v_s = \exp[-\text{const}(p - p_c)s^\sigma]$ and the simple exponential formula of Eqs. (25) and (26), $v_s = \exp[-\text{const}\,|\,p - p_c\,|^{1/\sigma}s]$ in common? In both cases, the function $v_s(p) = n_s(p)/n_s(p_c)$, which depends on the two variables s and $(p - p_c)$, is a function of the combination $|\,p - p_c\,|\,s^\sigma$ only. In the first case the function is an exponential of this combination; in the second it is an exponential of some power of this combination. Thus we may write in both cases:

$$v_s(p) = f(z) \qquad z = (p - p_c)s^\sigma$$

an equation supposed to be valid for p near p_c and large clusters. Inserting the usual law (Eq. (20)) at the critical point into this assumption we arrive at our final form:

$$n_s(p) = s^{-\tau}f[(p - p_c)s^\sigma] \qquad (p \to p_c,\ s \to \infty) \qquad (33)$$

The precise form of the scaling function $f = f(z)$ has to be determined by (computer) experiments and other numerical methods and is not predicted by our assumption. While this assumption replaces Eq. (25), our previous results (Eqs. (24), (29)–(32)) and our definition (Eq. (4)) remain valid. However, $f(z)$ nearly always turns out to approach a constant value for $|\,z\,| \ll 1$ (i.e. $s \ll s_\xi$), and to decay rather fast for $|\,z\,| \gg 1$. Thus, the role of $s_\xi \propto |\,p - p_c\,|^{-1/\sigma}$ as a cutoff and as a crossover size is maintained.

As we will discuss in more detail later, the assumption that there exists only one crossover size s_ξ is the main basis for a *single-variable scaling* like Eq. (33). The renormalization group theory (Chapter 4) gives some theoretical basis for this assumption, for space dimensions $d < 6$. For $d > 6$, it turns out that one needs additional lengths. For $6 < d < 8$, Harris and Lubensky (1981) derived a *two-variable scaling* behaviour of $n_s(p)$. For $d > 8$, however, these complicated arguments still reproduce the form (33). In fact, for $d > 8$ one reproduces the Bethe lattice results $\tau = 5/2$ and $\sigma = 1/2$. The Bethe lattice results for the moments, e.g. Eq. (31), hold for *all* $d > 6$. We shall say more on the cluster structure for $d > 6$ later. For simplicity we restrict our discussions from now on to $1 < d < 6$.

First let us see whether this new assumption solves the problems we had with the older form (Eq. (25)). Does the one-dimensional case of Eq. (1) now fit? The one-dimensional result after Eq. (26) was incompatible with that earlier assumption. But now we may take $\sigma = 1$, $z = (p - p_c)s^\sigma$, and rewrite this one-dimensional case as

$$n_s(p) = s^{-2}f(z) = s^{-2}z^2 \exp z \qquad (34)$$

valid again for p near unity and s large. Thus we see that now one dimension fits into the picture and corresponds to $\sigma = 1$ and $\tau = 2$. It is somewhat unusual since at $p = p_c$ there are no clusters left. Mathematically this effect results in

$f(0) = 0$, but we see no objections why that function cannot be zero at zero argument.

Can we now avoid the appearance of an infinite cluster even below p_c when we apply Eqs. (24) and (28)? The first part of Eq. (28) is still valid, and then we proceed as in the integrations of the preceding section, using $dz/ds = \sigma a/s$ and β from Eq. (29):

$$
\begin{aligned}
-P &= \sum_s [n_s(p) - n_s(p_c)]s \\
&= \int s^{1-\tau}[f(z) - f(0)]\,ds \\
&= |p - p_c|^{(\tau-2)/\sigma} \int |z|^{-1+(2-\tau)/\sigma}[f(z) - f(0)]\,dz/\sigma \\
&= (\beta + \gamma)|p - p_c|^\beta \int |z|^{-1-\beta}[f(z) - f(0)]\,dz
\end{aligned}
$$

Here the integration over $z = (p - p_c)s^\sigma$ goes from 0 to ∞ for $p > p_c$ and from 0 to $-\infty$ for $p < p_c$. Thus the mystery of the infinite cluster is solved. The scaling function $f(z)$ has to behave, for negative arguments, such that

$$\int |z|^{-1-\beta}[f(z) - f(0)]\,dz = 0$$

or

$$\int |z|^{-\beta}\left[\frac{df}{dz}\right]dz = 0$$

For positive arguments, that is above p_c, the corresponding integral should not vanish in order to give a non-zero strength of the infinite network. In order to give a vanishing integral below the percolation threshold, the function $f(z)$ has to be sometimes larger and sometimes smaller than $f(0)$ and cannot always increase if z increases from $-\infty$ (where f vanishes) to zero. Nature made it simple for us: $f(z)$ has only one maximum, and not many, for usual percolation problems. We call that value of $f(z)$ at this maximum f_{max}, and the negative value of z at this maximum is called z_{max}. Thus,

$$f(z_{max}) = f_{max} \qquad f(z) < f_{max} \quad \text{for } z \neq z_{max} \tag{35a}$$

For a fixed cluster size s, the cluster number n_s thus has a maximum at p_{max} below p_c, with

$$p_{max} = p_c + z_{max}s^{-\sigma} \tag{35b}$$

As a further test, before we go to numerical checks, we want to find out whether our assumption (Eq. (33)) leads to prohibited divergences in derivatives of cluster numbers with respect to p, as did Eq. (25). From what we have assumed so far we cannot exclude that possibility since, after all, Eq. (25) is a special case of our Eq. (33) with log $f \propto -z^{1/\sigma}$. Therefore we now assume in addition that $f(z)$ is an '*analytic*' function, which means very(!) roughly that all derivatives of $f(z)$ with respect to z are finite everywhere and in particular at $z = 0$. Since $dz/dp = s^\sigma$ that means also that all derivatives of $n_s(p)$ with respect to p remain finite at $p = p_c$, as they should. Thus the three problems mentioned at the end of the last section seem to be solved.

Finally, let us check whether the exponent γ of Eq. (30) can now be rederived:

$$S \propto \sum s^2 n_s$$
$$\propto \int s^{2-\tau} f(z) \, ds$$
$$= |p - p_c|^{(\tau-3)/\sigma} \int |z|^{-1+(3-\tau)/\sigma} f(z) \, dz/\sigma$$
$$\propto |p - p_c|^{-(3-\tau)/\sigma} = |p - p_c|^{-\gamma}$$

Thus Eq. (30) has been rederived. The total number of clusters can also be treated in a similar way, leading again to Eq. (32) and thus to the scaling law

$$2 - \alpha = 2\beta + \gamma$$

as already mentioned above. Of course, these purely theoretical consistency arguments do not yet prove assumption (33). There have been so-called renormalization group arguments in favour of (33) but mainly numerical evidence suggests assumption (33) to be correct.

The critical exponents like β and γ are important since they are 'universal', i.e. independent of the lattice structure and dependent only on the dimensionality. More precisely, in the scaling relation

$$n_s = q_0 s^{-\tau} f((p - p_c) q_1 s^{\sigma}), \tag{36}$$

the exponents τ and σ as well as the scaling function f are lattice independent, whereas only the proportionality factors q_0 and q_1 as well as p_c depend on the lattice details. If we now repeat the above calculation of $S = \Gamma(p_c - p)^{-\gamma}$ (for $p < p_c$) and $\Gamma'(p - p_c)^{-\gamma}$ (for $p > p_c$), and keep all the proportionality factors, then the 'amplitudes' Γ and Γ' are equal to $q_0 q_1^{-\gamma}/p_c$ multiplied by two integrals involving the 'universal' function f at positive and negative arguments.

Thus, the ratio

$$R = \frac{S(p_c - \varepsilon)}{S(p_c + \varepsilon)} = \frac{\Gamma}{\Gamma'}$$

i.e. the ratio of the amplitudes Γ on both sides, is also universal and equal to about 200 in two dimensions and 10 in three dimensions. Many other such amplitude ratios have been studied and also found to be universal, as reviewed by Privman *et al.* in Domb and Lebowitz (details in Further Reading of Chapter 1) (Vol. 14).

2.7. NUMERICAL TESTS

First, let us look at the exact cluster numbers of Section 2.3. For the square lattice we had $n_1 = p(1 - p)^4$ for the number of isolated sites and

$n_2 = 2p^2(1 - p)^6$ for the number of pairs. Do they have a maximum below p_c at fixed cluster size s, as the above argument requires? We find this maximum by setting the p-derivative of n_s equal to zero. For $s = 1$ we thus get

$$(1 - p)^4 - 4p(1 - p)^3 = 0$$

with the solution $p = 1/5$, whereas for $s = 2$ the same procedure leads to $p = 1/4$. In both cases, the position of the maximum is below the percolation threshold $p_c = 0 \cdot 5928$, and for the larger cluster the maximum (1/4) is closer to p_c than for the smaller cluster (1/5). This agreement with our theoretical expectation does not yet prove it, since our scaling assumption (Eq. (33)) is supposed to be valid only for large s. But if one plots $n_s(p)$ using the polynomials as calculated for various lattices by Sykes *et al.* until $s = 10$ to 20, one can determine the position of $p_{max}(s)$ for intermediate s and can show that p_{max} extrapolates for $s \to \infty$ to a value at least very close to p_c, just as Eq. (35) requires. In fact, similar methods have been used to determine p_c quite accurately.

One may also test Eq. (33) directly by calculating $v_s(p) = n_s(p)/n_s(p_c)$ from these exact polynomials and plotting the ratio versus $z = (p - p_c)s^\sigma$. Equation (33) then asserts that for different s the results all lie on the same curve $f = f(z)$. (Some people call this effect '*data collapsing*'.) Of course, in reality they do not all lie on the curve since even $s = 20$ is rather far from $s = \infty$, and Eq. (33) is only assumed to be valid for very large clusters. However, a rough confirmation of Eq. (33) has been obtained in this way.

The real strength of the exact polynomials for $n_s(p)$ lies in the determination of critical exponents like β, γ, and σ. For this purpose one is expanding all terms $(1 - p)^t$ by the binomial law and orders the result in powers of p. Thus, one arrives at a power series

$$M_k = \Sigma_i a_i p^i \tag{37}$$

for the moment M_k one is interested in, as defined in Eq. (31a). By looking at the radius of convergence of this series, for example for $k = 2$, one finds the percolation threshold p_c. Of course, this determination is inaccurate since only cluster numbers until $s = 10$ to 20 are known. Therefore only the first 10 to 20 terms of the series expansion (Eq. (37)) can be calculated. Suitable extrapolation methods have been developed, however, to find the threshold and the critical exponent with great accuracy from a limited number of expansion terms. (See for example the review of Gaunt and Guttman in vol. 3 of the Domb–Green series and of Adler *et al.* (1990).)

One of these extrapolations is called the *ratio method*. Consider, for example, $M_2(p)$, or $S(p)$. As we saw, $S(p)$ diverges as $(p_c - p)^{-\gamma}$ when p approaches p_c. The power series in p for $S(p)$ should thus have a finite radius of convergence, p_c. The function $(p_c - p)^{-\gamma}$ has the simple series expansion

$$(p_c - p)^{-\gamma} = p_c^{-\gamma}\left[1 + \frac{\gamma}{p_c} p + \frac{\gamma(\gamma + 1)}{2p_c^2} p^2 + \frac{\gamma(\gamma + 1)(\gamma + 2)}{6p_c^3} p^3 + \cdots\right]$$

Therefore, the ratio of two consecutive coefficients here is

$$\mu_i = \frac{a_i}{a_{i-1}} = \frac{\gamma + i - 1}{ip_c} = \frac{1}{p_c}\left(1 + \frac{\gamma - 1}{i}\right) \tag{38}$$

Plotted against $(1/i)$, this ratio exhibits a straight line, with slope $(\gamma - 1)/p_c$ and intercept $(1/p_c)$.

The actual series for $S(p)$ is not identical to that for $(p_c - p)^{-\gamma}$, since it may contain additional, less divergent, parts. However, for large i the series should be dominated by the most divergent parts, and therefore its ratio μ_i should approach the behaviour (38). Although some series have an oscillatory approach of μ_i to $(1/p_c)$, one often obtains good estimates for both p_c and γ with rather few terms.

More accurate estimates are often found using the *Dlog–Padé method*. Since we expect that $S = \Gamma(p_c - p)^{-\gamma}$ close to p_c, we also expect that $\log S = \log \Gamma - \gamma \log (p_c - p)$, or

$$\frac{d \log S}{dp} = \frac{\gamma}{p_c - p}$$

Given the series for $S(p)$, we can transform it into a series for $\log S$, using the expansion $\log(1 + x) = x - \frac{1}{2}x^2 + \frac{1}{3}x^3 - \frac{1}{4}x^4 \cdots$. We can then take a derivative, and find

$$\frac{d \log S}{dp} = \sum_{i=0}^{N} b_i p^i$$

We next try to represent this polynomial as the ratio of two other polynomials,

$$\sum_{i=0}^{N} b_i p^i = \frac{\sum_{i=0}^{L} c_i p^i}{\sum_{i=0}^{M} d_i p^i}$$

with $L + M = N$ and $d_0 = 1$. The righthand side has $(N + 1)$ unknown coefficients ($c_0, ..., c_L$ and $d_1, ..., d_M$). However, the requirement that the ratio on the righthand side has exactly the same Taylor expansion as the lefthand side is equivalent to the set of $(N + 1)$ linear equations (obtained by multiplying both sides by the denominator on the right)

$$\sum_{i=0}^{M} b_i d_{k-i} = c_k \qquad k = 0, ..., N$$

with $c_k = 0$ for $k > L$. Having solved these equations, we find several Padé approximants (depending on the choice of M and L) for d log S/dp. If they are to behave as $-\gamma/(p_c - p)$, then the polynomial in the denominator should vanish at p_c. We thus find the zeros of the denominator, i.e. the *poles* of the ratio. Usually only one of them will be a reasonable approximation for p_c (which must be real and obey $0 < p_c < 1$). Having chosen p_c, we replace the denominator by

$$\sum_{i=0}^{M} d_i p^i = (p_c - p) \sum_{i=0}^{M-1} e_i p^i$$

and our estimate for γ becomes

$$\gamma = - \frac{\displaystyle\sum_{i=0}^{L} c_i p_c^i}{\displaystyle\sum_{i=0}^{M-1} e_i p_c^i}$$

We next prepare a plot of the estimated p_c versus γ for all L and M. If many of these pairs concentrate around one particular point in the p_c–γ plane then we have a good estimate. Usually the best estimates are found when L and M are close to $N/2$.

For percolation, usually these series expansions give the most accurate estimates for the exponents. The determination of the threshold is often less accurate than that by the Monte Carlo method described in Chapter 4. Sometimes the best results are obtained from a Padé approximation which uses a Monte Carlo determination of p_c as input. Of course, when p_c is known exactly, that problem vanishes. Perhaps the greatest triumph of accurate series determinations for percolation was that Domb and Pearce (1976) estimated $\alpha = -0 \cdot 668 \pm 0 \cdot 004$ for triangular site percolation long before $\alpha = -2/3$ was first guessed and then nearly proved by theoretical methods (Nienhuis, 1982).

Note that the derivation of the series can be separated into two stages: one program is needed to evaluate the polynomials $n_s(p, d)$. This can be done for general dimensionality d. For example, for bond percolation on the hypercubic lattice one has

$$n_1(p, d) = dp(1 - p)^{4d-2} \qquad n_2(p, d) = d(2d - 1)p^2(1 - p)^{6d-4} \ldots$$

Once these are tabulated, we need a second program to evaluate the cluster property that we want to average. For M_k, this property was simply s^k. For another property A_s (e.g. the average time it takes the ant of Section 1.4 to go from any site to any other site on the cluster) we need a separate program. However, for many applications A_s depends only on the *topology* of the cluster (i.e. which site is neighbouring which) and not on the particular way the cluster sits on the d-dimensional lattice. Therefore A_s is calculated once for all d. The series for the average $\langle A \rangle$ for these topological quantities is then obtained via

$$\langle A \rangle = \sum_s \langle A_s \rangle n_s(p, d)$$

where $\langle A_s \rangle$ is the average over all clusters of size s. Analogous series, in powers of $(1 - p)$ instead of p, have also been derived and analysed, yielding information for $p > p_c$.

Now we come to the second method of numerically estimating properties of percolation, the *Monte Carlo* simulation. For thermal systems, the Metropolis algorithm is an old technique to simulate thermal fluctuations. However, for percolation the problem is usually much simpler since the system now is completely random, without memory effects. Thus, we may produce a picture like Fig. 2 by simply going through the lattice once and occupying each place randomly with probability p. After the production of the configuration one may analyse it by eye or by computer. We shift the latter problem to Appendix A where we describe how one can efficiently count the clusters in such a lattice or check whether the system is percolating. Here we concentrate on how to produce a configuration and what to do with the resulting cluster numbers.

Most computers have a built-in *random number* generator, and even programmable hand calculators often produce random numbers. These random numbers are generated neither at the roulette tables of Monte Carlo, where the name comes from, nor by students who failed in an examination, but by the electrons in the computer circuits. Simple arithmetic or logical operations produce in a completely predictable way a series of numbers which for the outsider, and the simulation, look quite random although they are not. The generation of such pseudorandom numbers is most easily explained on a 32-bit IBM computer, using FORTRAN language. Take IBM to be a large odd positive integer, called the 'seed' of this particular sequence of random numbers. Then a new odd positive and seemingly random integer can be produced (at least on IBM computers) by

IBM = IBM∗65539
IF(IBM.LT.0) IBM = IBM + 2147483647 + 1

The product of IBM and 65539 in general has more than ten decimal digits, that is more than 31 bits, and the computer loses the leading digits (bits) and stores only the last 32 bits in the storage place called IBM. The first of these 32 bits is interpreted as the sign of the number, and thus the product of two positive large integers may lead to a negative result. To avoid that accident one adds 2^{31} to IBM if IBM happened to come out negative (though one can also work successfully with negative IBM). The resulting IBM is an odd integer randomly distributed between zero and 2^{31}; it can be used as the input for the next random number to be generated, and so on. To get a real number between zero and unity one has to multiply the integer IBM by $0 \cdot 465566 \ \mathrm{E}{-9} = 2^{-31}$. This method of producing random numbers is often available as subroutine RANDU; but computation time is saved if the statements are written directly into the main program. At most 2^{30} different random numbers can be generated in this way, since there are no more odd integers between zero and 2^{31}. After at most 2^{30} steps, exactly the same sequence of pseudorandom numbers will be produced again; if you have bad luck, this happens after an even shorter period. In some applications, even with only several thousand lattice sites, problems occur which show that these pseudorandom numbers are not really random. Then a program published by Kirkpatrick and Stoll (1981) may help and is nearly as fast. On other computers, where this method may not work, you may use other subroutines, for example the function RANF( ) on Cray computers. In general the production of random numbers is an art and not a science; careful investigations should try different methods to check that the result does not depend on the method used.

As a simple introduction to random numbers and Monte Carlo simulation, let us calculate the number $\pi = 3 \cdot 14159 \ldots$, which is the area within a circle of radius unity. We take a random number which we call x and then another random number called y. Both random numbers are distributed with equal probability between zero and unity. Then we increase a counter, set initially to zero, by one unit if $x^2 + y^2 < 1$; otherwise we leave this counter unchanged. We repeat this computer experiment again and again, increasing the same counter if and only if the resulting $x^2 + y^2$ is smaller than unity. After, say, one million such pairs of random numbers have been used, our counter will be at about $10^6 \pi/4$. The pair (x, y) gives a point within the square of all points with $0 < x < 1$ and $0 < y < 1$. This point lies within the circle of radius unity with probability $\pi/4$. Thus counting the number of points with $x^2 + y^2 < 1$ measures the area of that circle.

How does one occupy a lattice randomly using these pseudorandom numbers? Let us assume we want to have in the array M(20, 20), which represents a 20×20 square lattice similar to that of Fig. 2, a zero stored for empty places and a one for occupied places, with concentration p. After using the above IBM random number generator we simply state

    M(I,K) = 0
    IF(IBM.LT.IP) M(I,K) = 1

where IP is the integer part of $p*2^{31}$. Thus, instead of normalizing the random integer IBM by a factor $1/2^{31}$ for each site M(I,K) we only multiply p by 2^{31} once. Then the integer IBM has a probability p of being smaller than IP, and has a probability $(1 - p)$ of being larger. Thus, the above statements fill the lattice with the desired probability.

In a finite lattice of, say, one million sites with concentration $p = 0 \cdot 3$ one does not occupy exactly 300 000 sites by this method; it should be several hundred more or less. Different techniques are necessary should one need exactly 300 000 occupied sites. If one merely wants to produce pictures where sites occupied for a lower p are also occupied for larger p, then one can do that easily, though not efficiently, by restarting with all sites empty for every new p, but always using the same initial seed IBM for the random number generator. Then for different p the same sequence of random numbers is used and if such a random number is smaller than $p = 0 \cdot 3$ (after normalization) it is also smaller than $p = 0 \cdot 4$. Here the fact that our random numbers are not really random has been used to our advantage, since now a site occupied for small p is automatically also occupied for large p.

Having produced a randomly occupied lattice we may check by eye whether or not it percolates, or we may count clusters. If it is done manually we will get such bad statistics that a meaningful test of the scaling hypothesis is hardly possible. If we do it on a computer we will get much better results. For $s = 1, 2$, up to 10 or 20 we can plot the results by hand, and then we will get tired and the statistics for n_s will get bad. But up to that size exact cluster numbers are available (see above), and there is no need to produce these clusters by Monte Carlo simulation. Therefore we have to concentrate on larger clusters up to size 1000. Then, as usual in computer work, we must protect ourselves against a flood of data in which we drown before we have analysed it. One way to do that is to combine different neighbouring cluster sizes in one *bin*. For example, one bit contains all clusters with 8 to 15 sites, the next bin corresponds to s between 16 and 31, then comes the interval 32 to 63, and so on, the bin size increasing exponentially. One may plot the result at the geometric mean of the two border sizes s, for example at $s = 45$ for the interval 32 to 63. Then even for very large lattices and very good statistics one does not have too much data to deal with.

Figure 12 shows the cluster numbers for a $95\,000 \times 95\,000$ triangular lattice at the exact $p = p_c = 1/2$, based on one run which took nearly 14 hours on a CDC Cyber 76 computer. We see a nice straight line in this log-log plot,

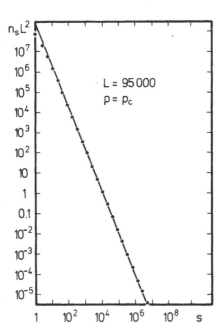

Fig. 12. High-quality data on cluster numbers at the percolation threshold, based on one simulation of a $95\,000 \times 95\,000$ triangular lattice. The slope of the straight line in this log-log plot gives $-\tau$, the Fisher exponent of Eq. (20). From Margolina *et al.* (1984); see also Rapaport (1985, 1992).

corresponding to the Fisher power law, Eq. (20). The exponent τ, as given by the negative slope of the straight line, seems to be very close to 2, as it should be. Theoretically, $\tau = 2 \cdot 055$. There are, however, two exceptions. For small s the data points fall below the line since the simple power laws of scaling theory are valid only for large s. And for very large s near 10^6, the data points seem to be too high, since the boundaries of the lattice cut the infinite cluster into several pieces, thus enlarging the number of clusters. We see that even a lattice size 95 000 is far from infinity. (Often one can reduce these boundary perturbations by working with *periodic boundary conditions*. For example, in a 20×20 lattice, the right neighbour of the rightmost site $M(i, 20)$ is taken as the leftmost site $M(i, 1)$ in the same row, and the topmost line $M(1, k)$ has the bottom line $M(20, k)$ as top neighbours. This was not done in the example of Fig. 12.)

Having tested the validity of scaling theory right at the percolation threshold, we now want to test Eq. (33) above and below p_c. Instead of giving high-quality data again as reviewed, for example, in Stauffer (1979) (see Further Reading, Chapter 1), we now take poor data which the reader may easily squeeze out of a medium-size computer overnight. Appendix A.3 gives a complete FORTRAN program and computer output for the simulation of one 500×500 lattice at concentrations of 38, 39, 40, ..., 62 per cent. A fast CDC 76 computer needed slightly more than one second for each concentration. We ignore the first four bins in the cluster size distribution, that is the sizes $1, 2-3, 4-7$, and $8-15$, since for such small clusters scaling is not good. Let us take the fifth bin, $16-31$, as an example. At $p = 0 \cdot 5$ we found 195 clusters in this size range, at $p = 0 \cdot 62$ only 4. Near $p = 0 \cdot 39$ is the concentration p_{max} for this size range, with 878 clusters observed, more than at lower or higher concentrations. We take the ratio $n_s(p = 0 \cdot 39)/n_s(p = 0 \cdot 50) = 878/195 = 4 \cdot 5$ and plot this number at $z = (p - p_c)s^\sigma = -0 \cdot 11 \times (22 \cdot 6^{0 \cdot 3956}) = -0 \cdot 378$, using $22 \cdot 6$ as the (geometric) average cluster size and $\sigma = 36/91$. The other data can be processed in a similar way and lead to Fig. 13.

We see in Fig. 13 that the different symbols, representing four different size ranges, all follow roughly the same bell-shaped curve. This is exactly what the scaling assumption (Eq. (33)) asserted—that all data points follow the same curve $f(z)$. Thus, within the very limited accuracy of this test run we have confirmed the scaling theory. Better data, for higher dimensions (Nakanishi and Stanley, 1980), confirm this validity with greater precision. Figure 13 also shows clearly the maximum below the percolation threshold, as opposed to the symmetric scaling function for the Bethe lattice.

If we therefore believe in scaling we may collect the present values of the various critical exponents from the literature. The above scaling laws relate them to each other and to other exponents to be introduced later. Therefore, if one exponent is not estimated directly with sufficient accuracy, we calculate it from other exponents. In this way, Table 2 gives d-dimensonal exponents as well as their 'classical' counterparts for the Bethe lattice. The two-dimensional exponents are believed to be exact on the basis of theoretical

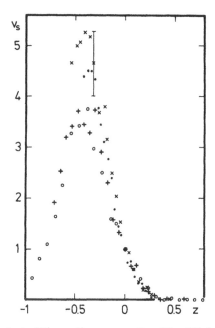

Fig. 13. Low-quality test of the scaling assumption (Eq. 33) for the cluster numbers, using the computer output in Appendix A. Different symbols correspond to different size ranges: dots to 16–31, crosses to 32–63; pluses to 64–127; squares to 128–255. Equation (33) requires all these ratios $v_s(p) = n_s(p)/n_s(p_c)$ to follow the same curve $f(z)$, where $z = (p - p_c)s^\sigma$. Within the strong scattering of the data that rule seems to be fulfilled.

analogies with thermal phase transitions, which are beyond the scope of this book. The 'striptease' method to be explained in Chapter 4 confirmed them with accuracies of the order of 10^{-4}. The higher-dimensional exponents are much less accurate.

Table 2 does not distinguish between different types of two-dimensional lattice, such as square, triangular or honeycomb lattice. All presently available evidence strongly suggests that the critical exponents as well as certain ratios like f_{max} depend only on the dimensionality of the lattice, but not on the lattice structure itself. In other words, if you have seen one two-dimensional lattice you have seen them all for these simple percolation problems. The same is true for d dimensions. Also, bond and site percolation have the same exponents. This simple fact, that critical exponents are independent of the lattice structure, is known in the trade under the complicated name of '*universality*' and also holds for thermal phase transitions. For each study of critical exponents it allows us to select that lattice for which our work is easiest.

Our computer output in Appendix A also gives the size 'INF' of the largest cluster and the second moment 'CHI' of the cluster size distribution

Table 2. Percolation exponents for $d = 2, 3, 4, 5, 6 - \varepsilon$ and in the Bethe lattice together with the page number defining the exponent. Rational numbers give (presumably) exact results, whereas those with a decimal fraction are numerical estimates.

Exponent	$d = 2$	$d = 3$	$d = 4$	$d = 5$	$d = 6 - \varepsilon$	Bethe	Page
α	$-2/3$	-0.62	-0.72	-0.86	$-1 + \varepsilon/7$	-1	39
β	$5/36$	0.41	0.64	0.84	$1 - \varepsilon/7$	1	37
γ	$43/18$	1.80	1.44	1.18	$1 + \varepsilon/7$	1	37
ν	$4/3$	0.88	0.68	0.57	$\frac{1}{2} + 5\varepsilon/84$	$1/2$	60
σ	$36/91$	0.45	0.48	0.49	$\frac{1}{2} + O(\varepsilon^2)$	$1/2$	35
τ	$187/91$	2.18	2.31	2.41	$\frac{5}{2} - 3\varepsilon/14$	$5/2$	33
$D(p = p_c)$	$91/48$	2.53	3.06	3.54	$4 - 10\varepsilon/21$	4	10
$D(p < p_c)$	1.56	2	$12/5$	2.8	–	4	62
$D(p > p_c)$	2	3	4	5	–	4	62
$\zeta(p < p_c)$	1	1	1	1	–	1	56
$\zeta(p > p_c)$	$1/2$	$2/3$	$3/4$	$4/5$	–	1	56
$\theta(p < p_c)$	1	$3/2$	1.9	2.2	–	$5/2$	54
$\theta(p > p_c)$	$5/4$	$-1/9$	$1/8$	$-449/450$	–	$5/2$	54
$f_{\max}$	5.0	1.6	1.4	1.1		1	42
μ	1.30	2.0	2.4	2.7	$3 - 5\varepsilon/21$	3	91
s	1.30	0.73	0.4	0.1_5		0	93
D_B	1.6	1.7_4	1.9	2.0	$2 + \varepsilon/21$	2	95
$D_{\min}(p = p_c)$	1.13	1.34	1.5	1.8	$2 - \varepsilon/6$	2	97
$D_{\min}(p < p_c)$	1.17	1.36	1.5		–	2	98
$D_{\max}(p = p_c)$	1.4	1.6	1.7	1.9	$2 - \varepsilon/42$	2	97

For the exponents at p_c, the Bethe lattice values are exact at $d \geqslant 6$. A dash means that 6 is not the upper critical dimension for the ε-expansion.

(excluding the largest cluster). The first is supposed to vanish at the threshold with an exponent β, whereas the second should diverge with the same exponent γ whether p_c is approached from below or above. We plot data points in Fig. 14; they do not seem to follow a straight line. Of course, for such a 'small' lattice one should not expect perfect agreement with theory since the boundaries cut smaller pieces off the largest cluster, etc. Finite-size scaling theory, Chapter 4, deals with such effects. One may repair part of these systematic errors due to the finite system size by working with an effective percolation threshold which differs from the true $p_c = 1/2$ but approaches the true threshold if the system size goes to infinity. Thus, we shift a trial value for the effective threshold until the exponent γ is the same above and below the threshold. Using all our data from $p = 0.4$ to $p = 0.6$ we find (via log-log paper, hand calculator, or personal computer) that for an effective threshold at 0.50827 the two exponents agree and have the value $\gamma = 2.39$, in excellent agreement with the desired $43/18$. Figure 14 shows these data too. By measuring the distance between the two parallel lines on the log-log plot for this shifted threshold value, one finds that the second moment of the cluster

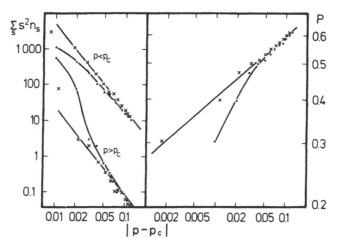

Fig. 14. Log-log plot of the strength P of the infinite cluster (right) and of the second moment M_2 of the cluster size distribution (left) for the triangular lattice, using the low-quality data of the computer output in Appendix A (500×500 triangular lattice). For $p_c = 1/2$, the exact value, the data are difficult to analyse (dots). If p_c is shifted to 0.5083 to take into account some finite size effects (crosses), we get reasonable straight lines with slopes close to the exponents found in more accurate studies.

size distribution is about 200 times larger below the threshold than at an equal distance above the (shifted) threshold, in agreement with determinations from larger lattices. This is the universal ratio R mentioned after Eq. (36). Using this same effective p_c and all data for the largest cluster from $p = 0.51$ to $p = 0.6$, we find the exponent β to be about 0.17, in reasonable agreement with the desired $5/36$. Thus, by suitable analysis one can squeeze out good results even from bad data. Also we see here that the usual formulae to estimate error bars for straight-line fits are unreliable, since they ignore the systematic errors due to finite system size (somehow corrected here), finite $p - p_c$, etc.

Besides the Monte Carlo method described here the reader can find other methods to produce clusters. For example, Leath (1976) starts from one site and then again and again adds, with probability p, a neighbour to the already existing cluster.

2.8. CLUSTER NUMBERS AWAY FROM p_c

Not everything in life is connected with critical phenomena near p_c. There are also interesting effects in the cluster numbers far away from the percolation threshold. It may be possible that some of their aspects are then approximated reasonably well by the Bethe lattice solution, or by effective medium theories. We concentrate here on (presumably) exact results for large clusters. In a way similar to a good politician, nature provides us with a compromise here: Sometimes it agrees with our Bethe lattice solution and sometimes it disagrees.

First let us argue why for concentrations below the percolation threshold the cluster numbers decay exponentially with cluster size s, that is

$$\log n_s \propto - s \qquad (s \to \infty, \, p < p_c) \tag{39}$$

We saw in Eq. (12) that the total number g_s of cluster configurations ('lattice animals') varies exponentially with s, that is

$$\log g_s \propto + s \qquad (s \to \infty)$$

apart from the less important contribution from the pre-exponential factor. Equation (13) has already explained why an analogous result should be valid for the cluster numbers, for small p, since then $n_s(p) = g_s p^s$. From this result we immediately get Eq. (39), as well as less important contributions, logarithmic in s, from the pre-exponential factor. Kunz and Souillard (1978) as well as Schwartz (1978) have shown that this exponential decay is valid for all p below some characteristic value p' which in turn is smaller than p_c. Numerical tests, as performed, for example, with the exact cluster numbers of Section 2.3, already support Eq. (39) for medium s.

However, there is a general belief, supported by renormalization group arguments to be discussed later, that Eq. (39) is valid for all p below p_c, in other words, $p' = p_c$. Thus, we may write this unproven result as

$$n_s(p < p_c) \propto s^{-\theta} \, \mathrm{const}^s \qquad (s \to \infty) \tag{40}$$

where we also use an exponent θ (see Eqs. (12) and (13)) for the pre-exponential factor. This exponent is thought to be the same for all p below p_c. Since Eq. (40) is supposed to be valid for all p below the threshold it is also valid for small p; therefore, everywhere θ equals its values 1 and 3/2 for two- and three-dimensional animals, respectively, as mentioned after Eq. (12). However, as mentioned after Eqs. (26) and (33), Eq. (40) holds only when $s \gg s_\xi$.

The behaviour above the percolation threshold is more interesting in that the dimensionality d enters the asymptotic decay law:

$$\log n_s(p > p_c) \propto - s^{1-1/d} \tag{41}$$

or

$$n_s(p > p_c) \propto s^{-\theta'} \exp(-Cs^{1-1/d}) \tag{42}$$

Again this law is supposed to be valid for large s only, i.e. $s \gg s_\xi$. Equation (41) was proved for all p above some p'' which is larger than p_c, and Eqs. (41) and (42) are thought to be valid for all p above the threshold for simple percolation problems.

The term $s^{2/3}$ (for $d = 3$ dimensions; the exponent is $(1 - 1/d)$ in d dimensions) in the exponential of Eq. (42) suggests a behaviour dominated by a *surface*, since the surface area varies as (volume)$^{(1-1/d)}$ (see Eq. (14)). Indeed, following Kunz and Souillard (1978), we can make this surface term plausible by the following argument. Let us look at the infinite network, to which above p_c a finite fraction of all lattice sites belongs. How much effort is needed to cut out from this network a finite, roughly spherical, cluster of radius r (in units of the lattice spacing)? The geometrical surface of this sphere is $4\pi r^2$. To transform the interior of this sphere into a finite cluster we have to cut all its connections with the outside infinite network. This can be done by making all sites on the surface of the sphere empty, that is less than about $4\pi r^2$ sites. The probability for a fixed configuration of s occupied and t empty sites generally is $p^s(1 - p)^t$, as Eq. (10) told us. To cut a finite sphere out of the infinite network, the change in both s and t is proportional to the surface area $4\pi r^2$, and the probability that such a cut occurs randomly varies therefore as $\exp(-\text{const } r^2)$. The number of large finite clusters is at least as large as the number of spherical clusters cut out randomly in this fashion. Therefore, n_s cannot be smaller than $\exp(-\text{const } r^2)$ in three dimensions, provided one has an infinite network present. In d dimensions we get analogously

$$n_s(p > p_c) > \exp(-\text{const } r^{d-1}) \propto \exp(-\text{const } s^{(1-1/d)})$$

since the number s of network sites in a (hyper)sphere of radius r varies as r^d. This result means that the cluster numbers cannot decay with a simple exponential as in Eq. (39), since that decay would be too fast. For large s, s is always larger than any term proportional to $s^{(1-1/d)}$.

This simple argument explains why Eq. (39) below the threshold is replaced by Eq. (41) above the threshold. From another inequality one can show that $|\log n_s|$ is indeed proportional to $s^{(1-1/d)}$ and not only smaller. Again this latter result is not rigorously proven for all p above p_c but widely believed to be valid in that whole range for simple percolation provided $s \gg s_\xi$. Numerical data, for example from the exact cluster numbers at intermediate s, confirm that Eq. (41) is a good approximation already for s near 10. (The exact exponent θ' in Eq. (42) is 5/4 in two and $-1/9$ in three dimensions according to Lubensky and McKane, 1981.) In the Bethe lattice we have no such difference between Eq. (39) and Eq. (41) but we also see why. The Bethe lattice corresponds to infinite d, and then (and only then) $(1 - 1/d)$ and 1 are identical. Unfortunately, these exactly known exponents like $(1 - 1/d)$ have not led to an exact solution for the three-dimensional percolation exponents near the threshold.

There is no contradiction between Eqs. (39)–(42), presumably exact away from p_c, and the scaling assumption (33), presumably true near p_c. For p close to p_c but s so large that $|z| = |p - p_c| s^\sigma$ is much larger than unity, both Eqs. (39)–(42) and Eq. (33) are expected to be valid. Therefore, the scaling function $f(z)$ in Eq. (33) must behave for large $|z|$ in such a way that Eq. (40) is fulfilled below and Eq. (42) is fulfilled above the threshold. For $p > p_c$, for example, we need

$$f(z) \propto z^{(\tau - \theta')/\sigma} \exp(-\text{const}' z^{(1-1/d)/\sigma})$$

to achieve that aim, with a simpler law below p_c. Monte Carlo data is in reasonable agreement with this seemingly complicated formula (see, for example, Stauffer (1979) cited in Chapter 1).

We may summarize the main result of this section with the help of an exponent ζ defined through

$$\log n_s \propto - s^\zeta \qquad (s \to \infty,\ p \text{ fixed}) \tag{43a}$$

Then

$$\zeta(p < p_c) = 1 \qquad \zeta(p > p_c) = 1 - 1/d \tag{43b}$$

In the next chapter we will investigate whether we also see this difference between above and below p_c (where a surface term only appears above p_c), in the structure of clusters. What this chapter has tried to do, in a somewhat unhistorical way, is to explain why percolation cluster numbers have to behave the way they do. Then we resorted to numerical tests to check that we were right. Historically it was much more a case of the opposite: Computers told us that we were wrong until our brains hit the right solution.

FURTHER READING

The King's College group (M.F. Sykes, D.S. Gaunt, M. Glen, H. Ruskin) published their full 'perimeter polynomials', Equation (10), in *J. Phys. A*, **9**, 87, 1705, 1899 (1976); **11**, 1369 (1978); **14**, 287 (1981).

The Metropolis method of Monte Carlo simulation is reviewed by K. Binder (ed.), *Monte Carlo Methods in Statistical Physics* (Heidelberg: Springer Verlag, 1986) and *Applications of the Monte Carlo Method in Statistical Physics* (Heidelberg: Springer Verlag, 1987).

Adler, J., *Physica A*. **171**, 453 (1991).
Adler, J., et al. *Phys. Rev. B*, **41**, 9183 (1990).
Domb, C. and Pearce, C.J., *J. Phys. A*. **9**, L 137 (1976).
Essam, J.W., Gaunt, D.S. and Guttmann, A. J., *J. Phys. A*, **11**, 1983 (1978).
Fisher, M.E., *Physics*, **3**, 255 (1967).
Flory, P.J., *J. Am. Chem. Soc.*, **63**, 3091 (1941).
Harris, A.B. and Lubensky, T.C., *J. Phys. Rev. B*, **24**, 2656 (1981).
Kirkpatrick, S. and Stoll, E.P., *J. Computational Phys.*, **40**, 517 (1981).
Kolb, M. and Herrmann, H.J., *J. Phys. A.*, **18**, L (1985).
Kunz, H. and Souillard, B., *J. Statist. Phys.*, **19**, 77 (1978).
Leath, P.L., *Phys. Rev. B*, **14**, 5046 (1976).
Lubensky, T.C. and McKane, A.J., *J. Phys. A*, **14**, L157 (1981).
Margolina, A., Djordjevic, Z., Stanley, H.E. and Stauffer, D., *Phys. Rev. B*, **28**, 1652 (1983).
Margolina, A., Nakanishi, H., Stauffer, D. and Stanley, H.E., *J. Phys. A*, **17**, 1683 (1984).
Mertens, S., *J. Stat. Phys.*, **58**, 1095 (1990).
Nakanishi, H. and Stanley, H.E., *Phys. Rev. B*, **22**, 2466 (1980).
Nienhuis, B., *J. Phys. A*, **15**, 199 (1982).
Parisi, G. and Sourlas, N., *Phys. Rev. Lett.*, **46**, 871 (1981).
Rapaport, D.C., *J. Phys. A*, **18**, L175 (1985) and *J. Stat. Ohys.*, **66**, (1992).
Redelmeier, D.H., *Discrete Math.*, **36**, 191 (1981) as analysed by Guttmann, A.J., *J. Phys. A*, **15**, 1987 (1982).
Schwartz, M., *Phys. Rev. B*, **18**, 2364 (1978).

CHAPTER 3
Cluster Structure

So far, we have looked only on the distribution of cluster sizes. We now turn to discuss the *geometry* of the clusters. We first look at the 'surface' of a cluster, i.e. its 'perimeter'. We then introduce the cluster's linear size, via its *radius*. In Section 1.3 we saw that the incipient infinite cluster has an internal *fractal geometry*, reflected by the dependence of its density on the length scale. We now discuss similar fractal relations between the radii of *finite* clusters at p_c and their masses (Eq. (48)). Scaling arguments are then presented to show that these results also hold for $p \neq p_c$, for length scales small compared with the correlation length ξ. For larger length scales one observes a *crossover* to different behaviours. Similar scaling arguments are then applied to quantify the description of Section 1.3 and to relate the fractal dimension D to other exponents (Eq. (54)). These discussions also introduce *hyperscaling*.

3.1. IS THE CLUSTER PERIMETER A REAL PERIMETER?

In Section 2.3 we introduced the '*perimeter*' t of a cluster, which is the number of empty sites neighbouring an occupied cluster site. We may call the size s of a cluster, the number of occupied sites, the *mass* of this cluster; then t is one of the quantities which define the structure of this mass. The word perimeter suggests that it is some sort of surface, similar to the perimeter of a circle, which is $2\pi \times$ radius and thus proportional to the square root of the 'mass' (area) of πr^2 of the circle. Thus one might expect, at first sight, that t is also proportional to $s^{(1-1/d)}$ in d dimensions, analogously to Eq. (14). The aim of this section is to show that this is not so.

We only have to look at Fig. 2 to see that the infinite cluster for concentrations p above the percolation threshold p_c has some holes in its interior. Each of these holes gives a contribution to the perimeter. If we have one hole for, say, every thirty sites we have a perimeter proportional to the number of sites in the infinite network. For a very large but finite cluster one may expect the same behaviour as for the infinite network and thus also a perimeter proportional to the number of sites in the cluster. Thus,

$$t \propto s \qquad (s \to \infty)$$

57

seems plausible according to these arguments. If correct, this quantity t is not a quantity which may be identified directly with a cluster surface.

Do you want a proof? Leath (1976) (as cited in Chapter 2) has given one. First we have to define the average perimeter t_s of a cluster containing s sites, for Eq. (10) tells us that different clusters with the same mass s have different perimeters. We take

$$t_s = \Sigma_t \frac{t n_{st}}{n_s}$$

where

$$n_{st} = g_{st} p^s (1 - p)^t$$

is the average number of s-clusters having t perimeter sites each, as is obvious from Eq. (10); of course, $\Sigma_t n_{st}$ gives n_s. If we differentiate the quantity n_{st} with respect to p we get

$$\frac{dn_s}{dp} = \Sigma_t g_{st} [s p^{s-1}(1 - p)^t - p^s t (1 - p)^{t-1}]$$

which leads to

$$t_s = s \frac{1 - p}{p} - (1 - p) \frac{d \ln(n_s)}{dp} \tag{44a}$$

Now we insert what we learned in the last section, that $\ln(n_s) = -Cs^\zeta$ apart from terms varying less strongly with s, with a p-dependent factor C. Thus also $d \ln(n_s)/dp$ varies as s^ζ, and

$$t_s = s \frac{1 - p}{p} + \text{const } s^\zeta \qquad (s \to \infty) \tag{44b}$$

We see from Eq. (44b) that for sufficiently large clusters the perimeter t_s is always proportional to the mass s. Thus, the perimeter is not a surface in the usual sense. Even deep in the interior of the cluster one has perimeter sites, just as holes in a Swiss cheese (or water in a Norwegian fjord) prevent the solid cheese (or earth) from filling the space completely. Only the second term in Eq. (44b) may correspond to a usual surface contribution, since for $p > p_c$ we have $\zeta = (1 - 1/d)$ from Eq. (43), giving a perimeter contribution proportional to the usual surface.

You may think that the perimeter does give a real surface if one restricts it to the external perimeter, that is to those sites which are connected by a chain of empty sites to the space far away from the cluster. Indeed, in the Swiss cheese the interior holes have no connection to the outside air and thus do not correspond to the external perimeter. However, for the simple cubic lattice at least, we can see easily that even this external perimeter varies as the volume s, and not as a surface $\propto s^{2/3}$. Let us take p between $0 \cdot 4$ and $0 \cdot 6$. Thus p lies between $p_c = 0 \cdot 312$ and $(1 - p_c)$. Now not only the occupied sites (concentration p) percolate through the lattice but also the empty sites (concentration $(1 - p)$). Nearly every occupied site is part of the infinite network of occupied sites, and nearly every empty site belongs to the infinite network of connected empty sites. Every large cluster of occupied sites is penetrated by a web of empty sites connected with the outside. Thus, even the external perimeter will be proportional to s, and not to $s^{2/3}$. In two dimensions, this argument is not valid. At p_c, every cluster has many internal 'holes', which cannot be reached from the outside. One can thus separate 'external' and 'internal' perimeter sites, and only their total combined number obeys Eq. (44). As we shall show below (Section 6.4), there are several ways to define the 'external' perimeters, and these have interesting fractal geometries.

3.2. CLUSTER RADIUS AND FRACTAL DIMENSION

While we have seen that 'surfaces' are difficult to define, the 'radius' of a complicated object is much easier to study. Polymer scientists have always had to deal with objects more complicated than a straight line, a square or a sphere. They usually define a 'radius of gyration' R_s for a complicated polymer through

$$R_s^2 = \Sigma_{i=1}^s \frac{|r_i - r_0|^2}{s} \tag{45a}$$

where

$$r_0 = \Sigma_{i=1}^s \frac{r_i}{s} \tag{45b}$$

is the position of the *centre of mass* of the polymer, and r_i is the position of the ith atom in the polymer. We now use the same definition for our percolation problem, replacing 'polymer' by 'cluster' and 'atom' by 'occupied site'. If we average over all clusters having a given size s, the average of the squared radii is denoted as R_s^2. If we turn a two-dimensional cluster around an axis through its centre of mass and perpendicular to the cluster, then the kinetic energy and angular momentum of this rotation is the same as if all sites were on a ring of radius R centred about the axis. Therefore, such radii are called 'gyration' radii. We may also relate R_s to the average distance between two cluster sites:

$$2R_s^2 = \Sigma_{ij} \frac{|r_i - r_j|^2}{s^2} \tag{45c}$$

as one can derive easily after putting the origin of the coordinates into the cluster centre-of-mass: $r_0 = 0$.

The *correlation function* $g(r)$ is the probability that a site at distance r from an occupied site is also occupied and belongs to the same cluster (see end of Section 2.2). The average number of sites to which an occupied site at the origin is connected is therefore $\Sigma g(r)$, the sum running over all lattice sites *r*. On the other hand, this average number equals $\Sigma_s s^2 n_s/p$, since $n_s s/p$ is the probability that an occupied site belongs to an *s*-cluster, that is to a cluster containing mutually connected sites. Thus,

$$pS = \Sigma_s \, s^2 n_s = p \, \Sigma_r \, g(r) \qquad (p < p_c) \qquad (46)$$

The second moment of the cluster size distribution equals the sum over the correlation function (apart from an uninteresting factor *p*). Above p_c this relation is also valid if the contribution from the infinite cluster is subtracted. This amounts to replacing $g(r)$ everywhere by $g(r) - P^2$. Such a relation between mean cluster size (or second moment) and correlation function has already been mentioned in Eq. (9). The word 'connectivity' function is also used for our $g(r)$.

We define the correlation or connectivity length ξ as some average distance of two sites belonging to the same cluster:

$$\xi^2 = \frac{\Sigma_r \, r^2 g(r)}{\Sigma_r \, g(r)} \qquad (47a)$$

Since for a given cluster, $2R_s^2$ is the average squared distance between two cluster sites, since a site belongs with probability $n_s s$ to an *s*-cluster, and since it is then connected to *s* sites, the corresponding average over $2R_s^2$ gives the squared correlation length:

$$\xi^2 = \frac{2 \, \Sigma_s \, R_s^2 s^2 n_s}{\Sigma \, s^2 n_s} \qquad (47b)$$

Thus, apart from numerical factors, the correlation length is the radius of those clusters which give the main contribution to the second moment of the cluster size distribution near the percolation threshold. We expect ξ to diverge as *p* approaches p_c, as

$$\xi \propto |\, p - p_c \,|^{-\nu} \qquad (47c)$$

For two-dimensional percolation, plausible but not rigorous arguments give $\nu = 4/3$, in excellent agreement with numerical results. In three dimensions, ν is somewhat smaller than $0 \cdot 9$, whereas for Bethe lattices one has $\nu = 1/2$, analogous to numerous mean-field theories for thermal phase transitions. See also the discussion on cluster structure at high dimensions, towards the end of Section 5.3.

We have seen in Chapter 2 that many quantities diverge at the percolation threshold. Most of these quantities involve sums over all cluster sizes *s*; their main contribution comes from *s* of the order of $|\, p - p_c \,|^{-1/\sigma}$ (Eq. (33)). Now we see that the correlation length, which is also one of these quantities (Eq. (47b)), is simply the radius of those clusters which contribute mainly to the divergences. This effect is the foundation of scaling theory. There is one and only one length ξ dominating the critical behaviour. In contrast to politics, this dictatorial principle works quite successfully for all sorts of critical phenomena in two and three dimensions, not only for percolation.

However, the dictatorship principle does not always work even in critical phenomena. If a sum over all cluster sizes s is not diverging at the critical point, then the main contribution to this sum does not come from clusters with radii of the order of the correlation length. Instead the main contribution comes from small s near, say, 10. For example, the sum $\Sigma n_s s$ equals p as long as p is not larger than p_c. Thus it does not diverge at the critical point. At $p = p_c$ the dictatorship principle asserts that the main contribution comes from $s = \infty$; but for the sum over $s^{1-\tau}$ the main contribution comes from rather small s. Then one has to be very careful in the evaluation of critical exponents. Let us take as an example three different definitions of an average squared radius:

$$\frac{\Sigma_s R_s^2 n_s s^2}{\Sigma_s n_s s^2} \qquad \frac{\Sigma_s R_s^2 n_s s}{\Sigma_s n_s s} \qquad \Sigma_s \frac{R_s^2 n_s}{\Sigma_s n_s}$$

Here the first expression is our definition of the squared correlation length and diverges with the exponent 2ν. In the second expression the denominator remains finite at the threshold whereas the numerator diverges with the exponent $(2\nu - \beta)$ (Eq. (31b), see also Eq. (105) later). In the third expression, in d dimensions, for $d > 2$ both the numerator and the denominator remain finite, and the exponent is 0. Thus, depending on the type of averaging the particular situation requires, the critical exponent varies appreciably. In short, do not rely on dictators. (Polymer scientists call the first expression a z-average, the second a weight average, and the third a number average over the squared cluster radius.)

We now want to find out how the radius R_s varies with s at the percolation threshold. Our discussion in Section 1.3 showed that the largest cluster at p_c has a fractal behaviour, with $M \propto L^D$. It is thus natural to assume that also $s \propto R_s^D$, with the same D. We can now relate D to ν. We start by substituting

$$R_s \propto s^{1/D} \qquad (p = p_c, \ s \gg 1) \tag{48}$$

into Eq. (47b).

The denominator of Eq. (47b) is the kth moment of the cluster size distribution with $k = 2$ (Eq. (31a)), and diverges with the exponent $\gamma = (3 - \tau)/\sigma$ (Eq. (31b)). If near p_c the radius R_s varies as $s^{1/D}$, the numerator of Eq. (47b) is a moment with $k = 2 + 2/D$ and thus diverges with the exponent $(3 - \tau + 2/D)/\sigma$ according to the same Eq. (31). Thus their ratio diverges with the exponent $2/(D\sigma)$. This exponent should equal 2ν according to Eq. (47c). Thus,

$$\frac{1}{D} = \sigma\nu \tag{49}$$

As mentioned in Section 1.3, this fractal dimension D is $91/48 = 1 \cdot 896$ in two and about $2 \cdot 5$ in three dimensions. Thus the finite clusters at the percolation threshold are fractals in the sense that their fractal dimension D is smaller than their lattice dimension d.

For Bethe lattices, Zimm and Stockmayer (1949) showed that $D = 4$ is the same for all p, not only at p_c. Can we expect also for three-dimensional percolation that D is the same above, at and below the threshold? We cannot. Imagine we have p very close to unity. Then Eq. (10) tells us that only those clusters with the smallest perimeter are important in averages over all cluster configurations. The smallest perimeter, for a cluster of $s = L^3$ sites on a simple cubic lattice, is obtained for configurations having no holes at all in their interior; their perimeter is $6L^2$ and their average radius is of the order of L. Thus R_s is proportional to $s^{1/3}$ for p close to unity, and not to $s^{\sigma \nu} = s^{0.4}$ as for $p = p_c$. We have seen in Section 2.8 that one should expect the same type of asympototic behaviour for all p above p_c (Eq. (43)). Thus we also expect this $s \propto R_s^3$ law to be valid for the radius for all p larger than p_c: $D = 3$. In d dimensions we thus have

$$D = d \qquad (p > p_c) \tag{50}$$

which shows that D is not the same as at the threshold. Clusters above p_c are not fractals but 'normal' objects with $D = d$, provided $s \gg s_\xi$.

Percolation theory supports equal rights for clusters above and below p_c; if above p_c they have the freedom to deviate from Eq. (49), those below p_c have the same freedom. Again, one expects the same D for all p below p_c. In the limit $p \to 0$, all different perimeters in Eq. (10) get the same weight, which means that the cluster radius is now the average radius of all animal configurations of the given size s. (All animals are equal, none of them are more equal than others.) Unfortunately, no general exact solution is known for animal radii, but in three dimensions we have $D = 2$ exactly (Parisi and Sourlas, as cited after Chapter 2). In two dimensions, $1/D$ is about 0.641 from numerical estimates, and as mentioned above, $D = 4$ in the Bethe lattice. Thus the animals as well as the percolation clusters below p_c are again fractals, but with a fractal dimension D smaller than that at the percolation threshold. The table of exponents in Chapter 2 summarized the situation for D.

As we have mentioned several times, many of the results quoted as power laws are only *asymptotic*, i.e. they are valid only for very small $(p - p_c)$ or for very large s. This also applies to Eq. (48), which should hold only for $s \gg 1$. At finite s, there appear *corrections* to the asymptotic behaviour, and these may involve new exponents:

$$s = AR_s^D(1 + aR_s^{-\Omega} + \text{smaller corrections}) \tag{51}$$

Usually it is sufficient to consider a single correction term. Normally, we suggest checking relations like Eq. (48), for Monte Carlo data, on a log-log plot. From Eq. (51), we find

$$\log s = \log A + D \log R_s + \log(1 + aR_s^{-\Omega})$$

The *local* slope of this function, near a cluster size s, is

$$\frac{d \log s}{d \log R_s} = D - \frac{\Omega a R_s^{-\Omega}}{1 + a R_s^{-\Omega}} \simeq D - \Omega a R_s^{-\Omega}$$

This local slope may be considered as an *effective fractal dimension*, D_{eff}. If we measure only over a finite, narrow range of sizes, the measured local slope may mislead us into identifying a wrong value D_{eff} for the asymptotic D. A better technique would involve finding the local slope (e.g. by fitting a straight line to a set of data of width Δs around s), and then plotting it versus some negative power of R_s, attempting a few exponents Ω. The correct choice of Ω will yield asymptotically a straight line of D_{eff} versus $R_s^{-\Omega}$, and will have an intercept D. Indeed, this procedure has been applied successfully to such plots both for percolation and for lattice animals, with rather straight lines when Ω was chosen as 1. These simulations then confirm the above expression for D_{eff}. Similar tricks are useful for other asymptotic power laws.

Equation (51) should hold *at p_c*. We abbreviate the inverse fractal dimension $1/D$ at, below and above p_c by ρ, ρ' and ρ''. When we move away from p_c, we might expect a gradual crossover from Eq. (48) with $1/D = \rho$ for $1 \ll s \ll s_\xi$ to either ρ' ($p < p_c$) or $\rho'' = 1/d (p > p_c)$ for $s \gg s_\xi$. The assumption of a single-variable scaling would then imply a scaling function of the form

$$R_s = s^\rho h[(p - p_c)s^\sigma] \tag{52}$$

with $h(x)$ approaching a constant for $|x| \ll 1$, $x^{(\rho' - \rho)/\sigma}$ for $x \ll -1$ and $x^{(\rho'' - \rho)/\sigma}$ for $x \gg 1$. Plots of $\log R_s$ versus $\log s$ at $p \neq p_c$ will thus also produce effective slopes. At finite $|p - p_c|$ we have a finite $s_\xi \propto |p - p_c|^{-1/\sigma}$, and the slope will gradually change from ρ to ρ' or to ρ'' as s increases beyond s_ξ.

Although we are not aware at present of detailed numerical tests of the scaling law (52), there is no reason not to believe it. Equation (52) as well as Eq. (33) are manifestations of single variable scaling: a function $f(x, y)$ of two variables turns out to have the form

$$f(x, y) = x^{-A} g(y/x^B)$$

implying that all the information falls on a *single curve* if we plot $x^A f$ versus y/x^B. In later chapters we relate this scaling behaviour to the role of ξ as the only relevant length, and we explain it using the modern theory of the *renormalization group*.

As we saw, Eq. (52) contains three different asymptotic fractal dimensions, valid above, at and below p_c. What does that mean for the structure of the clusters? If we define an average density as the ratio s/R_s^d of cluster mass to cluster volume, then above p_c with $s \propto R_s^d$ one has an average density independent of the cluster size for large clusters as mentioned in Section 1.3. This density equals the strength P of the infinite cluster since the interior of a very large cluster should not be different from that of the infinite network. Below and at p_c, the strength P of the infinite cluster is zero, and therefore the average density approaches zero if the cluster mass s goes to infinity. Franke has given the density profiles of large but finite clusters, that is the probability that a site at distance r from the cluster centre-of-mass belongs to that cluster. For p above p_c, the interior region of high densities near P is separated from the outside of the cluster (zero-density profile) by a relatively narrow surface layer (Franke, 1981, 1982); below and at p_c this surface layer has spread over the whole interior of the cluster. Thus above p_c one has a rather narrow surface whereas below and at p_c the surface extends over the whole volume. Now it is no longer surprising that the exponent ζ for the asymptotic decay of cluster numbers (Eq. (43)), has the surface value $(1 - 1/d)$ above p_c but is unity (surface proportional to volume) below p_c. Thus we see that the sites of finite clusters in percolation theory have some human traits. Above the threshold, the sites stay together like workers in a trade union and thus they achieve higher densities. Below the threshold they seem to prefer some distance from each other, they scatter over a large volume, the links between them can be broken more easily, just as with non-unionized people, and thus they achieve only low densities. In this sense you reach a percolation threshold if you enter a union.

3.3. ANOTHER VIEW ON SCALING

As noted above, the critical behaviour is dominated by the single diverging length, ξ. From Eq. (47b), ξ represents the radius of the clusters which give the main contribution to the mean cluster size and similar properties. From Eqs. (48) and (49), the size of these clusters is

$$s_\xi \propto \xi^D \propto (p - p_c)^{-D\nu} \propto (p - p_c)^{-1/\sigma}$$

This is exactly the cluster size that dominated the moments of the mass distribution, see Eq. (31). In fact, $s_\xi \propto 1/c$ appeared as a cutoff on this mass distribution: Eqs (25, 33) have the power law behaviour $n_s \propto s^{-\tau}$ for $s \ll s_\xi$ and are exponentially small for $s \gg s_\xi$. The behaviour for $s \ll s_\xi$, or for $R_s \ll \xi$, is indistinguishable from that at p_c. We can thus identify ξ as the *crossover length*, separating the 'critical' behaviour $n_s \propto s^{-\tau}$ and $R_s \propto s^{1/D}$ from the different behaviours described in Sections 2.8 and 3.2.

Crossover phenomena are very common in statistical physics. They are always associated with a *length scale*, like ξ. For length scales which are much smaller than ξ one may ignore the existence of a finite ξ, and the behaviour is the same as that found when ξ is infinite (i.e. $p = p_c$ in our example). In the absence of any basic length scale to use as a 'measuring stick', all the relevant functions become power laws. The power law is the only function that does not require another length. For example, for $p \neq p_c$ we saw that the correlation function $g(r)$ essentially decays exponentially, like $\exp(-r/\xi)$ (see

Eq. (7)). Such an exponential dependence (which also identifies ξ as the cutoff length, beyond which it is not probable to find pairs of points on the same cluster) was possible only because we could construct the *dimensionless* ratio (r/ξ); the function e^x requires that its argument x be dimensionless (a length, like r, has dimensions, i.e. it is measured with units such as metres or kilometres). At p_c, or when $r \ll \xi$, $g(r)$ must become a power law. Indeed, one finds that at p_c

$$g(r) \propto r^{-(d-2+\eta)}$$

and the (relatively small) exponent η is related to other exponents via the scaling relation $d - 2 + \eta = 2\beta/\nu$.

3.4. THE INFINITE CLUSTER AT THE THRESHOLD

Is there or is there not an infinite cluster present at $p = p_c$? We know that there is one for p above the threshold, and there is none for p below the threshold. What does Nature do at the border line?

First, we have to clarify what we mean by an infinite network, since real systems are always finite. We may call a cluster infinite if it connects the top line (top plane) with the bottom line (or plane). In this case, in a computer simulation of large lattices at $p = p_c$, a finite fraction, for example one-half, of all lattices have an infinite cluster in this sense, and the rest do not. Thus the answer to whether an infinite cluster is present is simply a 'perhaps'.

Instead we may look at the largest cluster in the finite system (without periodic boundary conditions). Of course, even for p far below p_c the system has a largest cluster. But only for p above p_c is the size of this largest cluster of the order of the system size. Generally we ask: How does the size of the largest cluster increase with L in a system with L^d sites? It is determined by $n_s L^d = 1$; far below p_c, from Eq. (13) we thus conclude that $s \propto \ln L/\ln(p \times \text{const})$, whereas at p_c, s follows a power law. Thus at p_c it would seem reasonable that the largest cluster will have a radius of the order of the system length L: $R_s \propto L$. Since at $p = p_c$ we have $R_s \propto s^{1/D}$, the condition for the largest size is simply $L \propto s^{1/D}$, very similar to Eq. (48). Thus the infinite cluster at p_c (which sometimes is called the incipient infinite cluster) also is a fractal in our sense and has the same fractal dimension D as large finite clusters at the threshold. Above p_c, the mass s of the infinite cluster increases as L^d which means that is is no longer a fractal but $D = d$ as for the finite clusters. Below p_c, the fractal dimension of the largest cluster is zero (corresponding to a very weak, that is logarithmic, increase with L), in contrast to the fractal dimension of the finite clusters which are animal-like with $D = 2$ in three dimensions.

Figure 15 shows results in two dimensions for lattices containing up to 10^{10} sites. Apart from fluctuations we see a simple straight line in this plot of $\log s$ versus $\log L$ when L is not too small. The slope of this line is close to

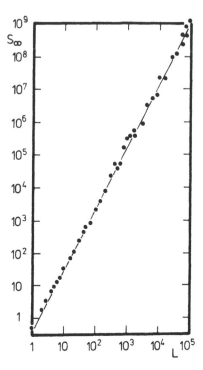

Fig. 15.   Monte Carlo data for the size of the largest cluster at the site percolation threshold $p = p_c = 1/2$ of the triangular lattice, as a function of the linear dimension L of the lattice. The slope of this log-log plot for large sizes gives the fractal dimension $D = 91/48 \approx 1 \cdot 9$.

the theoretical value $D = 91/48$. Thus now we have a more quantitative answer to how large is 'infinite' for the incipient infinite cluster.

In this picture we took $p_c = 1/2$ as is known exactly for the triangular lattice. But even if p_c is not known exactly and if one takes a p_c slightly too high, one still observes for the largest cluster a mass s proportional to L^D, as long as L is much smaller than the correlation length ξ. (At the exact threshold, ξ is infinite, and this condition is always fulfilled.) On the other hand, as explained in Section 1.3, if L is much larger than ξ, the mass of the largest cluster will be PL^d, with P again being the strength of the infinite cluster, proportional to $(p - p_c)^\beta$. If L is of the order of ξ these two expressions PL^d and L^D are of the same order (assuming again that only one length ξ is dominating critical behaviour):

$$PL^d = \text{const } L^D \quad \text{at} \quad L = \xi \propto (p - p_c)^{-\nu}$$

Thus

$$\beta - d\nu = -\nu D$$

or

$$D = d - \beta/\nu$$

Together with $\nu D = 1/\sigma = \beta + \gamma$ and the other scaling laws we get

$$d\nu = \gamma + 2\beta = 2 - \alpha = \frac{\tau - 1}{\sigma} \tag{53}$$

Often this scaling law where the dimensionality d enters is called 'hyper-scaling', and is also used for thermal phase transitions.

As we have noted, the Bethe lattice corresponds to a very large dimension d. Using the Bethe lattice values $\sigma = 1/2$, $\tau = 5/2$ and $\nu = 1/2$, it is obvious that Eq. (53) breaks down for large d. The only dimension at which it holds, with these exponents, is $d = 6$. Equation (53) is also believed to work for all $d < 6$. $d_u = 6$ is thus identified as the *upper critical dimension*. We shall show later (Section 5.3 and Appendix B) that for all $d > 6$ one has $\sigma = 1/2$, $\tau = 5/2$ and $\nu = 1/2$, but hyperscaling breaks down and the fractal dimension maintains its value at $d = 6$, i.e. $D = 4$. Unless specifically stated, we restrict ourselves to $d < 6$.

Equation (48), as well as the discussion leading to Eq. (52), were based on a study of the dependence of the *total* mass of a finite cluster, s, on its linear size, R_s. In fact, the fractal behaviour also contains information on the *internal* structure of clusters. This was demonstrated in Section 1.3 by looking at the largest clusters shown in Fig. 2. When $p \gg p_c$, we noted qualitatively that the cluster looks rather *homogeneous*, with typical holes of sizes 1–3 lattice sites. As one moves down towards p_c, these holes grow larger, and one encounters a distribution of their sizes. Since large holes in the larger cluster contain finite clusters, it is not surprising that the linear sizes of these large holes are connected to those of the finite clusters, with a cutoff of order ξ. Thus, the cluster is homogeneous on length scales much larger than ξ, and rather ramified, with holes on all scales smaller than ξ. In fact, the cluster looks *fractal* on scales smaller than ξ. A quantitative way to study this was described in Section 1.3: choose a point on the cluster and then draw squares of variable sizes L around it. The mass of the cluster within each such box is denoted by $M(L)$, and the corresponding density is $\rho(L) = M(L)/L^d$ (with $d = 2$ for the analysis in two dimensions). The results of measurements of $\rho(L)$ (which we described qualitatively in Section 1.3), are shown in Fig. 16.

At $p - p_c = 0.035$, the double-logarithmic plot shows a constant slope of -0.1 for $L < 10$, implying that in this range $\rho(L) \propto L^{-0.1}$ and $M(L) \propto L^{1.9}$. For $L > 10$, the curve has zero slope, indicating a constant density, i.e. a homogeneous mass distribution. The crossover length ξ (equal to about 10 for $p - p_c = 0.035$) moves up as p approaches p_c, and the constant density for $L > \xi$ moves down. In the homogeneous regime it is natural to identify ρ with the fraction of sites which belong to the infinite network, i.e. its 'strength'. Indeed, Kapitulnik *et al.* (1984) showed that the crossover length ξ and the

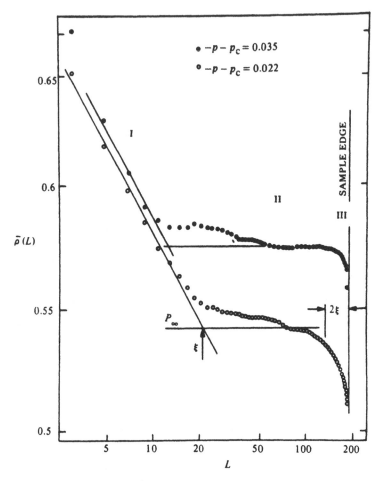

Fig. 16. Density of sites which are connected to a point on the largest percolating cluster on a square lattice, at $p - p_c = 0 \cdot 035$ (solid circles) and $p - p_c = 0 \cdot 022$ (open circles), within a box of size L around an occupied site. The slope for $L < \xi$ is $D - d = -0 \cdot 1$, and the plateau for $L > \xi$ is $P(p)$. From Kapitulnik *et al.* (1984).

plateau P behave as $\xi \propto (p - p_c)^{-\nu}$ and $P \propto (p - p_c)^\beta$, with $\nu \simeq 1 \cdot 33$ and $\beta \simeq 0 \cdot 14$.

Figure 16 demonstrates that although $P(p)$ is the *average* probability of an occupied site to be on the infinite cluster, the *density* of this cluster (relative to one of its points) is not uniform for lengths $L < \xi$. If we want to calculate physical properties of the cluster, we have to consider the detailed geometrical structure of the cluster on these length scales.

Since the cluster has a constant density for $L > \xi$, it is natural to divide the system into boxes of linear size ξ. In d dimensions, the total volume L^d will be divided into $(L/\xi)^d$ boxes. Since the cluster inside each of these boxes, of size ξ^d, has a mass of order ξ^D, the total mass of the cluster is given by

$$M(L, \xi) \propto \begin{cases} L^D & L < \xi \\ \xi^D (L/\xi)^d & L > \xi \end{cases} \tag{54a}$$

For $L > \xi$, $M \propto \xi^{D-d} L^d \propto P L^d$, hence $P \propto (p - p_c)^\beta \propto \xi^{D-d}$ and

$$D = d - \beta/\nu \tag{54b}$$

as obtained before Eq. (53). This hyperscaling law works only for $d < 6$.

Stated differently, we may say that for $L \ll \xi$, M depends only on L, and therefore must be a power of L, $M \propto L^D$. For L of order ξ or larger, additional dependence on L must appear only through the scaled ratio (L/ξ) (i.e. ξ is our only 'measuring stick'). Thus, we may write a scaling form

$$M(L, \xi) = L^D m(L/\xi) \tag{55}$$

For $L \gg \xi$, we expect that the L dependence of M becomes L^d, hence for $x \gg 1$ we expect that $m(x) \propto x^{d-D}$, yielding again the above result.

Note that Eq. (55) is very similar to Eq. (52), except that here we discuss the mass *within* a section of the largest cluster, which we denote by M, whereas in Eq. (52) we consider the mass of the *whole* cluster, s. The same logic, of dividing R_s into parts of size ξ, may be used to justify Eq. (52).

FURTHER READING

Franke, H., *Z. Physik B*, **40**, 61 (1981); **45**, 247 (1982), *Phys. Rev. B*, **25**, 2040 (1982).
Kapitulnik, A., Aharony, A., Deutscher, G. and Stauffer, D., *J. Phys. A*, **16**, L 269 (1984).
Zimm, B.H. and Stockmayer, W.H., *J. Chem. Phys.*, **17**, 1301 (1949).

CHAPTER 4
Finite-Size Scaling and the Renormalization Group

In 1971, K.G. Wilson published the first renormalization group treatment of critical phenomena and was honoured a decade later by the Nobel prize for physics (though in the cumulative author index of the journal at that time the articles were forgotten). It is an attempt to justify the scaling assumptions made earlier, and to calculate the critical exponents entering these scaling assumptions. Historically, it was first applied to thermal phase transitions and only afterwards to percolation; also initially it dealt with fluctuations in Fourier space (as function of wave vector) and only later moved to real space (where everything depends on distances). Ignoring this history, we will concentrate on real-space renormalization of percolation (sometimes also called position space renormalization). This seems the simplest way to introduce renormalization ideas into percolation theory (Reynolds *et al.*, 1980) since the method becomes for large lattices equivalent to finite-size scaling. Thus we will first explain finite-size scaling and then go to renormalization techniques.

We start this chapter with a description of finite size scaling, first for the dependence of properties calculated at p_c (Eq. (56)) and then for the shift in the threshold p_c (Eq. (59)) as function of the system's size. We then describe several examples of small cell renormalization group algorithms. The renormalization group idea yields some basis for the scaling hypotheses presented before. We then discuss computer large cell applications of the renormalization group, and end with a geometric interpretation of the correlation length exponent ν. Fourier space renormalization group for percolation (Harris *et al.*, 1975) is only very briefly mentioned in Appendix B.3. More details are given by Reynolds *et al.* (1980).

4.1. FINITE-SIZE SCALING

How do the various quantities of interest behave near the percolation threshold in a large but *finite* lattice? We have already touched on this general question several times in this book. In Section 1.3, Fig. 16 and Eq. (54) we saw that the mass of points connected to a point on the largest cluster within

70

a box of size L scales as L^D for $L \ll \xi$ and as $(p - p_c)^\beta L^d$ for $L \gg \xi$ (and $p > p_c$). These results were summarized in the scaling form of Eq. (55). These facts imply that the *density* $P(L, \xi)$ behaves as $(p - p_c)^\beta$ for $L \gg \xi$, but as $L^{-\beta/\nu}$ for $L \ll \xi$. Indeed, Fig. 15 showed that for sufficiently large samples at p_c we have the accurate power law $M \propto L^D$. We expect similar rules for other quantities. If a quantity X is predicted to scale as $|p - p_c|^{-x}$ for sizes $L \gg \xi$, then we expect it to obey the general scaling law

$$X(L, \xi) = \xi^{x/\nu} x_1(L/\xi) \propto \begin{cases} \xi^{x/\nu} & L \gg \xi \\ L^{x/\nu} & L \ll \xi \end{cases} \tag{56a}$$

or

$$X(L, p) = (p - p_c)^{-x} x_2((p - p_c)L^{1/\nu}) \tag{56b}$$

Studies of X as function of the system size L at p_c thus yield the exponent x/ν. Often this method gives more accurate results for x/ν than separate Monte Carlo studies for x and ν.

As a simple demonstration of this rule, recall percolation in one dimension (Section 2.2). For a finite sample, the mean cluster size S is an analytic polynomial in p. When $L \to \infty$, we saw that $S = (1 + p)/(1 - p) \propto \xi$ (Eq. (8)). On the other hand, if we set $p = 1$ at finite L, then there is clearly only one cluster, of size L. Thus $S = L$, confirming the one-dimensional result $x/\nu = 1$. If both L and ξ are very large but finite, one must repeat the derivation of Eq. (5) with a finite sum in the geometrical series $\Sigma_s p^s$, and one can see that indeed $S = \xi f(L/\xi)$, such that $f(x \gg 1) \to$ constant and $f(x \ll 1) \propto x$ (in this calculation, which we recommend as an exercise, you should be careful in calculating the probability of the clusters which neighbour the edges. For example, $n_L = p^L$ and $n_{L-1} = 2p^{L-1}(1 - p)$ and not as given in Eq. (1).)

We have also exhibited finite size results in Fig. 14. Since the data there were taken from rather small lattices, it was found that the data fit better the expected laws of $P \propto (p - p_c)^\beta$ and $S \propto (p_c - p)^{-\gamma}$ with a *shifted value* of p_c. We devote the rest of this section to explaining that effect.

How does one identify p_c from simulations on a *finite* sample? Since the sample is finite, there is a finite probability of finding a spanning cluster at *any* finite concentration. In the one-dimensional example, the probability of finding such a cluster is $\Pi = p^L = e^{-L/\xi}$. Thus, for a given L there is a probability larger than $1/e$ of finding a spanning cluster if $\xi > L$, i.e. $(1 - p) < 1/L$.

For an infinite sample we expect that $\Pi = 1$ above p_c and $\Pi = 0$ below p_c. Indeed, our one-dimensional result approaches this limit (with $\Pi = 1$ at $p = p_c = 1$ and $\Pi = 0$ for $p < 1$) when $L \to \infty$. Since Π is expected to approach the step function when $L \to \infty$, we might define an effective threshold p_c at the concentration where $\Pi = 1/e$ (the choice $\Pi = 1/2$, or $\Pi = 1/3$, is just as good). This effective threshold p_{eff} should approach the true p_c when $L \to \infty$. Indeed, in the one-dimensional example we saw exactly that $\xi(p_{\text{eff}}) = L$, hence $p_c - p_{\text{eff}} \propto 1/L$.

Let us now apply these ideas to higher dimensions, when p_c is smaller than unity. Let $\Pi(p, L)$ be the probability Π that a lattice of linear dimension L percolates at concentration p. (We define a lattice as percolating if at least one cluster connects the top line or plane with the bottom line or plane.) In an infinite system, we have $\Pi = 1$ above and $\Pi = 0$ below p_c. The quantity $d\Pi/dp$ gives the probability (divided by the small interval dp), that the lattice starts to percolate if the concentration is increased from p to $p + dp$. Since in infinite systems $\Pi = 1$ for all p above p_c, the critical exponent of Π is zero, and the analogue of Eq. (56) in a finite system is

$$\Pi = \Phi[(p - p_c)L^{1/\nu}] \tag{57a}$$

for large L close to p_c. The scaling function Φ increases from 0 to 1 if its argument increases from $-\infty$ (far below threshold) to $+\infty$ (far above threshold). The derivative gives

$$\frac{d\Pi}{dp} = L^{1/\nu}\Phi'[(p - p_c)L^{1/\nu}] \tag{57b}$$

(For $L \to \infty$, this derivative approaches a *delta function*, as physics students call it, or a delta distribution as mathematicians insist on calling it. Ordinary people simply call it a very narrow and high peak.) Figure 17 shows schematically how Π and $d\Pi/dp$ behave for medium and large lattices, in agreement with numerical studies.

The average concentration p_{av} at which, for the first time, a percolating

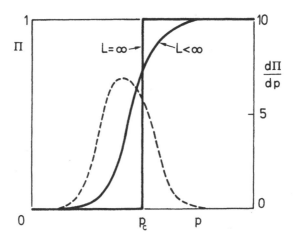

Fig. 17.   Variation of the probability Π (solid lines) that a cluster is spanning the whole system, for medium and large system sizes. The dashed lines give $d\Pi/dp$, proportional to the probability that at concentration p a spanning cluster starts to appear. The width of the transition region or peak varies as $L^{1/\nu}$. (Schematic only.)

cluster connects top and bottom of the cluster is defined as

$$p_{av} = \int p\left(\frac{d\Pi}{dp}\right) dp \tag{58}$$

the integral here and later runs from $p = 0$ to $p = 1$. (Note $\int(d\Pi/dp)\, dp = \Pi(1) - \Pi(0) = 1$.) Since $d\Pi/dp$ is basically the probability that at concentration p such a spanning cluster appears for the first time, we can determine p_{av} by making numerous Monte Carlo experiments for the same L and check when the system percolates for the first time when we slowly fill up the sites. For large lattices it is practical to achieve this by the following approximation. First take $p = 1/2$ and check whether the lattice percolates. If it does, decrease p by $1/4$, otherwise increase p by $1/4$. Then check again whether the lattice percolates; decrease p by $1/8$ if it does and increase p by $1/8$ if it does not. Repeat this division until one has determined with sufficient accuracy the concentration at which the first spanning cluster appears. After every change of p the random number generator has to be reset at its original value in order that in the following simulation most of the sites previously occupied (empty) will be occupied (empty) again. After about ten such iterations the onset of percolation is known with an accuracy sufficient for many purposes, but this value is only the value for this particular sequence of random numbers. Now we have to repeat the Monte Carlo simulation again and again and get, in general, each time a different concentration at which for the first time a cluster connects top and bottom. Averaging the observed onset concentration over all these different sequences of simulations we get an estimate for p_{av}. (Instead of checking for a spanning cluster one may also determine a different percolation threshold as the position of the maximum in the second moment of the cluster size distribution. For large systems these thresholds converge to the same limit p_c as those described here.)

How does this effective percolation threshold p_{av} for one system size L approach the asymptotic value p_c for infinite systems? From Eqs. (57) and (58) we find that, unless $\Phi'(z)$ is symmetric in z,

$$p_{av} - p_c \propto L^{-1/\nu} \tag{59}$$

with the proportionality constant being $\int z\Phi'(z)\, dz$. (In special cases in two dimension the proportionality constant may be zero since then $d\Pi/dp$ is completely symmetric about $z = 0$. In those cases p_{av} approaches p_c faster. See Ziff 1992). This variation of p_{av} with system size L is one way to determine the critical exponent ν: one plots p_{av} versus $L^{-1/\nu}$ for various trial values of ν and selects the value for ν which gives the best straight line for large L. (If one has lots of data and a suitable fitting program one can also determine ν, β and p_c as those parameters which best fulfill the finite-size scaling assumption of Eq. (56).) In our one-dimensional example, $d\Pi/dp = Lp^{L-1}$. Hence $p_{av} = L/(L + 1)$ and $1 - p_{av} = 1/(L + 1)$, in agreement with the above.

Not only p_{av} approaches p_c as $L^{-1/\nu}$. If we also define an effective percolation threshold for finite lattices as that point where the curve $P(p)$ has an inflexion point (maximum of dP/dp), this L-dependent threshold approaches the true p_c as $L^{-1/\nu}$, as Eq. (56) tells us immediately. By analogy, the maximum of the mean cluster size $S(p)$ gives us an effective threshold value which approaches the true threshold as $L^{-1/\nu}$. In other words, Eq. (59) is valid for every reasonable definition of a percolation threshold for finite large systems, not just for p_{av}. Only the proportionality constant is different for different definitions of the onset of percolation. This explains Fig. 14.

We mentioned above that the exponent χ/ν can be determined from $X \propto L^{\chi/\nu}$ by simulating a system exactly at $p = p_c$ for different L. We now see that we may also simulate it at some suitably defined size-dependent threshold, like p_{av}. For then the argument $z = (p - p_c)L^{1/\nu}$ in Eq. (56) is a constant, instead of being zero, and the proportionality for $X(L)$ is not affected.

All these remarks are also valid for thermal critical phenomena in two and three dimensions (Fisher, 1971) and were used there successfully before they were applied to percolation. One has only to replace P by the magnetization, and $p - p_c$ by $T_c - T$, if one deals with the ferromagnetic Curie point.

The width Δ of the transition region between small and large probabilities Π of a spanning cluster can be defined very simply as the difference between the concentration where Π is $0 \cdot 1$ and the concentration where Π is $0 \cdot 9$. One may also take $0 \cdot 2$ and $0 \cdot 8$, or $0 \cdot 1$ and $0 \cdot 8$, as suitable numbers to be used in the definition of this width. Equation (57a) then tells us that this width Δ, independent of the details of its definition, varies for large L as $L^{-1/\nu}$. A less arbitrary way to define the width is

$$\Delta^2 = \int (p - p_{av})^2 \left(\frac{d\Pi}{dp}\right) dp \qquad (60a)$$

Thus Δ is the root mean square deviation of the thresholds actually observed, from their average value. For averages $\langle x \rangle$ of fluctuating quantities x, one can easily show quite generally that the average of $(x - \langle x \rangle)^2$ equals $\langle x^2 \rangle - \langle x \rangle^2$. Thus one can easily determine Δ from a series of Monte Carlo evaluations of the actual onset p of percolation by summing the values of p as well as p^2 for each sequence of random numbers. At the end one can then calculate

$$\Delta^2 = \langle p^2 \rangle - \langle p \rangle^2$$

and determine the correlation length exponent ν from

$$\Delta \propto L^{-1/\nu} \qquad (60b)$$

(You can check that this holds for our one-dimensional example.) If the data are not accurate enough to allow for a precise determination of ν, and if ν is not known from other sources, we can still determine p_c by combining

Eq. (60b) with Eq. (59), when the latter holds:

$$p_{av} - p_c \propto \Delta \tag{60c}$$

This means we plot the observed thresholds p_{av} versus the observed widths Δ, and extrapolate to the intercept p_c by letting $\Delta \to 0$. No knowledge of ν is required here.

This method works particularly well for percolation, where it was introduced by Levinshtein *et al.* (1975). We will mention numerical results in a later section, where we will show the close connection with renormalization group techniques. Here we simply warn the reader that one needs at least hundreds of simulations to get the exponent ν accurately from Eq. (60), whereas far fewer are needed to estimate p_c reliably from Eq. (59). Thus, determinations of the threshold should be done with relatively few simulations of large lattices whereas one needs many more simulations for the determination of the correlation length exponent. Therefore, because of demands on computer time, these have to be done for smaller systems.

4.2. SMALL CELL RENORMALIZATION

The basic idea of renormalization is *self-similarity* at the critical point. What does this mean? We saw in Eq. (56) that the crucial question of finite-size scaling is the question whether the system length L is larger or smaller than the correlation length $\xi \propto |p - p_c|^{-\nu}$. We saw in Eqs. (33) and (52) that clusters can be separated into two main groups: those with mass s larger than $|p - p_c|^{-1/\sigma}$, and those with smaller s. For small clusters or small systems, one kind of power law is valid, for example $P \propto L^{-\beta/\nu}$, and for large clusters or systems another power law holds, for example $P \propto (p - p_c)^{\beta}$. In other words, all clusters or systems smaller (in linear dimension) than the correlation length ξ are similar to each other in an averaged sense, as long as they contain many sites. This similarity breaks down for large sizes of the order of ξ as well as for small sizes of the order of the distance a between nearest neighbours on the lattice. Right at the percolation threshold the correlation length is infinite; thus all large clusters or lattices are similar to each other. If we take out a medium size part of a bigger lattice, then both this part and the bigger lattice are still much smaller than ξ at $p = p_c$, and thus similar to each other in an average sense.

Although we suggested before that Fig. 2 demonstrates that similarity holds for distances between a and ξ, a critical reader would not really find such a similarity. Only if one averages over many pictures can a computer find such similarity. Figure 15 may be regarded as a much better 'proof'. If different lattice sizes were not similar to each other at the threshold one should not be able to observe a simple power law as seen by the straight line in Fig. 15. The similarity idea as a foundation of thermal critical phenomena and scaling goes back to the 1960s (see the review of Kadanoff *et al.*, 1967) and leads to Wilson's first renormalization theory.

In real-space renormalization, we replace a cell of sites by a single super-site, provided that the linear dimension b of the cell is much smaller than ξ. Of course we lose information if, say, 16 sites of a 4×4 cell in the square lattice are replaced by a single super-site. But if scaling relies on the fact that all cells of size b are similar to each other, then perhaps we should get a good critical exponent out from this approximation where we renormalize a whole cell of, say, b^d sites into a single super-site. Quantitatively, such a renormalization of cells to sites requires a certain rule governing how this is to be done; moreover the concentration p' of occupied super-sites will in general be different from that of the original sites. Only right at the critical point, where self-similarity is valid, do we have $p' = p = p_c$. In general we know that the correlation length ξ limits the validity of similarity; thus the limit ξ is the same in both the original lattice and the renormalized lattice of super-sites. If in the original lattice we have $\xi = \text{const} \, | \, p - p_c |^{-\nu}$ then in the renormalized lattice, with lattice constant b, we have $\xi' = \text{const} \, | \, p' - p_c |^{-\nu}$, with the same proportionality constant and the same critical exponent ν provided that both $| \, p - p_c |$ and $| \, p' - p_c |$ remain very small. However, the new lattice has a new lattice constant b, and ξ' is measured in these units; $\xi' = \xi/b$. Thus

$$b \, | \, p' - p_c |^{-\nu} = | \, p - p_c |^{-\nu} \tag{61a}$$

is the basic equation of real-space renormalization. Taking the logarithm of both sides we arrive at

$$\frac{1}{\nu} = \frac{\log [(p' - p_c)/(p - p_c)]}{\log b} = \frac{\log \lambda}{\log b} \tag{61b}$$

where $\lambda = (p' - p_c)/(p - p_c) = \mathrm{d}p'/\mathrm{d}p$ at $p = p_c$. Often $1/\nu$ is abbreviated as y or y_T or y_p in renormalization publications. In summary, we renormalize a cell of size b into a single super-site; to keep the real quantity ξ unchanged in this renormalization, we also have to renormalize p into p'.

To demonstrate the idea, let us return again to the one-dimensional case. Suppose we group the sites on the line into cells, with b sites in each cell. A cell will carry the connectivity from one of its ends to the other only if all of its sites are occupied. We can thus replace the cell by a new super-site, and identify the occupation probability of this super-site as

$$p' = p^b \tag{62}$$

If we start with $p = 1$, this yields also $p' = 1$. The point $p^* = 1$ is thus identified as a *fixed point* of this *renormalization group transformation*. A fixed point remains fixed under renormalization. If we start with $p < 1$, then Eq. (62) will yield $p' < p$. If we repeat this procedure many times, the renormalized concentration will approach zero. Indeed, $p = 0$ is also a fixed point, which we may associate with lattice animals.

At the fixed point $p = 1$, p' is the same as p. However, if we calculate

ξ in units of the distance between our new super-sites, we have $\xi' = \xi/b$. Since $p' = p$, we should also have $\xi' = \xi$. These two equations are consistent only if $\xi = \infty$ (or zero). We thus identify the fixed point $p^* = 1$ with the percolation threshold p_c.

We can now return to Eq. (62). From it,

$$\lambda = \frac{dp'}{dp} = bp^{b-1} = b \text{ at } p = 1$$

Therefore, Eq. (62) yields

$$\frac{1}{\nu} = \frac{\log \lambda}{\log b} = \frac{\log b}{\log b} = 1$$

in agreement with Eq. (7). The renormalization group turns out to be exact in one dimension.

The situation is less obvious in higher dimensions. As in the example above, we need to identify p' with the probability of the cell to connect opposite lines or planes. There exist many ways to choose the cells and to identify p', and none yields simple exact results. In this section we present two rather accurate and simple examples, one for site percolation on the triangular lattice and the other for bond percolation on the square lattice.

On the triangular lattice, each triangle contains three sites at its corners, and we place the super-site into the centre of the triangle. Figure 18 shows that we only renormalize suitable triangles in such a way that each original site belongs to exactly one triangle. We now ask for the probability p' of such a

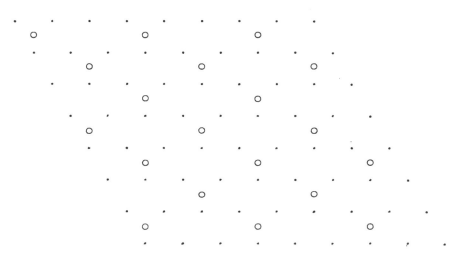

Fig. 18. Real-space renormalization of a triangular lattice. The circles denote the super-sites, each representing three different original sites. The super-sites again form a triangular lattice.

super-site belonging to a renormalized triangle being occupied if every original site is occupied with probability p. The super-site is occupied if a spanning cluster exists. In our triangle this is the case if either all three sites are occupied (probability p^3), or if two neighbouring sites are occupied and thus connect two opposite ends of the triangle. The latter case can be realized in three ways (depending on which of the three possible pairs is occupied), each of which has the probability $p^2(1 - p)$, see Fig. 18. Thus our renormalized probability is (Reynolds *et al.*, 1980)

$$p' = p^3 + 3p^2(1 - p) \qquad (63)$$

A plot of this function $p'(p)$ has some similarity with the function $\Pi(p)$ depicted in Fig. 17.

Right at the critical point we should have complete similarity: $p' = p = p^*$. In our case, the equation $p' = p$, with p' from Eq. (63), has three solutions:

$$p^* = 0, 1/2 \text{ and } 1$$

The first (zero) and the last (unity) solution are quite trivial and exist also for different lattices and dimensionalities; we are interested only in the non-trivial fixed point $p^* = 1/2$. This fixed point agrees exactly with the known critical point p_c of the triangular lattice, a first indication that the renormalization idea might be correct, after all. Now we expand Eq. (63) about this fixed point:

$$p' = p^* + \lambda(p - p^*) + O(p - p^*)^2$$

with

$$\lambda = dp'/dp = 6p - 6p^2 = 3/2 \qquad \text{at} \quad p = p^* = 1/2$$

In our particular lattice we have $b^2 = 3$ since in the plane three old sites form one super-site (see Fig. 18). Thus Eq. (61b) yields

$$\nu = \frac{\log(3^{1/2})}{\log(3/2)} = 1 \cdot 355$$

This result is an excellent approximation to the presumably exact $\nu = 4/3$ in two dimensions, confirming that renormalization group works.

This concludes our example for site percolation. We now present another approximate renormalization group scheme, which has turned out to be very accurate for *bond* percolation on the square lattice (Reynolds *et al.*, 1977; Bernasconi, 1978). In this scheme, one replaces the 2×2 cell, with eight bonds, as shown by Fig. 19(a), into a supercell, which has only two bonds (Fig. 19(b)). The two new bonds represent connectivities in the horizontal and vertical directions. If we consider only the horizontal direction, then we can ignore the two 'dangling' bonds DG and EH, and the connectivity from left to right is determined by the five bonds AB, BC, BE, DE and EF. Figure 19(c) shows generic configurations of these bonds, with their corresponding prob-

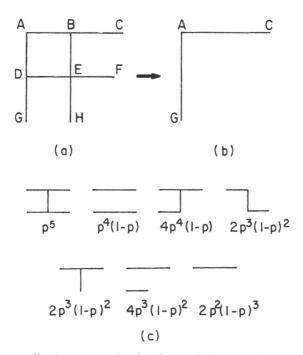

Fig. 19. Renormalization group for bond percolation on the square lattice. (a) Original 2 × 2 cell of eight bonds. (b) Renormalized cell, with two bonds. (c) Generic parts of original cell which contribute to the renormalized bond AC, with their corresponding probabilities.

abilities. The two edges AD and CF will thus be connected with probability

$$p' = p^5 + 5p^4(1 - p) + 8p^3(1 - p)^2 + 2p^2(1 - p)^3 = 2p^5 - 5p^4 + 2p^3 + 2p^2$$

(64)

where the different terms correspond to connecting configurations with 5, 4, 3 and 2 bonds. It is easy to check that

$$p^* = \frac{1}{2} \qquad b = 2 \qquad \text{and} \qquad \lambda = \frac{dp'}{dp}\bigg|_{p=p^*} = \frac{13}{8}$$

Thus $1/\nu = \ln \lambda / \ln b \approx 0 \cdot 700$, and $\nu \approx 1 \cdot 428$. Again, the value of $p^* = p_c$ is exact and the value of ν is quite close to the exact value of 4/3.

The fact that Eqs. (63) and (64) give the exact values of p_c, and good estimates for ν, need not mislead you. In fact, both equations are approximate. If you are interested mainly in working tools, you may now accept the fact that these two approximate small cell schemes work, and jump to the next section. In the remainder of the present section we give some technical feeling for the nature of the approximation, on the triangular case, and on the general ways to improve it. Figure 20(a) shows two disconnected clusters, which are combined into one cluster in the renormalized picture. To preserve

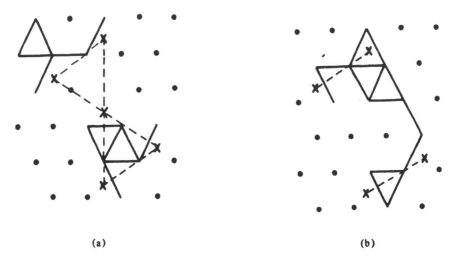

(a)   (b)

Fig. 20. Examples of renormalized group configurations, using the rules of Fig. 18. Old occupied sites are connected by full lines. New occupied sites are connected by broken lines. (a) Two disconnected clusters which become connected. (b) A connected cluster which splits into two disconnected clusters.

the identity of clusters, we should, for example, introduce a new bond concentration variable, x, which will indicate the probability that two neighbouring renormalized sites are connected. Similarly, Fig. 20(b) shows a portion of a single connected cluster, which splits into two disconnected ones in the renormalized picture. This maps the original site problem into the site–bond problem, which we mentioned towards the end of Section 2.1. If we restrict ourselves to the two parameters p (site occupation) and x (bond occupation), then the renormalized pair p' and x' 'flow' in the p–x plane as shown schematically in Fig. 21 (Nakanishi and Reynolds, 1979). All the points on the thick line, which separate the region with an infinite cluster (near $p = x = 1$) from that with finite clusters, have an infinite correlation length. These points flow to the non-trivial fixed point (p^*, x^*). Thus, every initial point which is close to this critical line will first flow towards (p^*, x^*), and then flow away from it. The critical exponent ν, related with the rescaling of ξ (Eq. (61a)), is thus determined from the linearized recursion relations near (p^*, x^*).

The site–bond model is not the end of the renormalization story. In general, further iterations will generate more parameters, e.g. next nearest neighbours site and bond concentrations, probabilities for three bond nodes, and so on. One thus 'flows' into a multi-dimensional parameter space. The only effective way we know to cope with this is to truncate the list of these parameters, and work within some approximate space.

Even in that generalized space, one normally finds only one non-trivial fixed point, and the vicinity of that fixed point determines the critical exponent ν for *all* possible initial parameters near percolation. This is the reason

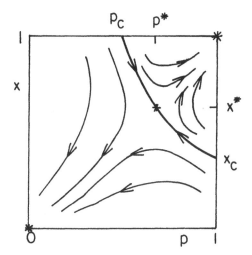

Fig. 21. Schematic renormalization group flow of the site–bond percolation problem. The stars denote the fixed points. The thick line consists of points which flow to the critical fixed point (p^*, x^*), all have an infinite correlation length. x_c and p_c are the pure bond and site percolation thresholds.

for the *universality* of the critical exponents: the renormalization gets rid of minor local differences, and maps all critical problems to the same vicinity of the fixed point.

4.3. SCALING REVISITED

What does the fixed point really mean? As we noted, $\xi(p^*) = \infty$. On the other hand, we identified p^* by requiring that the problem remain invariant under the grouping of sites in cells into 'super-sites', and changing the length scale by a factor b. To get an intuitive feeling for this invariance, look at Fig. 22(a). It shows a simulation of site percolation on the triangular lattice at $p_c = 1/2$. The sites on the largest cluster, which connect between the boundaries of the finite sample, are emphasized by showing the bonds which connect them. Figure 22(b) then shows the super-sites, obtained using the procedure described above. Qualitatively, Fig. 22(b) cannot be distinguished from a piece of Fig. 22(a). It is impossible to tell from the picture at what level of iteration, or magnification, or 'coarse graining', the true pictures were taken. This is a qualitative manifestation of *self-similarity* in a random sense.

The mass of the spanning cluster in Fig. 22(b) is roughly smaller by a factor b^D than that of the same cluster in Fig. 22(a). This can be expressed mathematically by the relation

$$M(L) = b^D M(L/b)$$

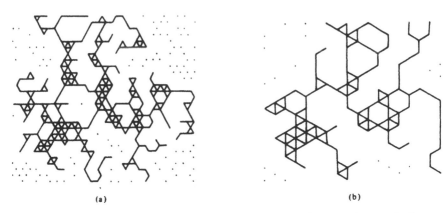

Fig. 22. Site percolation on a triangular lattice, at $p_c = 1/2$. The sites on the largest connected cluster are emphasized by the connecting bonds. (a) Original simulation. (b) Renormalized version, with triangular sites being occupied by the rule explained in Fig. 18 and Eq. (63).

After l iterations, this becomes

$$M(L) = b^{lD} M(L/b^l)$$

(Of course, this will become accurate if we average $M(L)$ and $M(L/b)$ over many samples. Fluctuations of individual samples around this rule are said to represent 'lacunarity'.) The only solution of this functional equation is the power law function $M(L) \propto L^D$, obtained by setting $b^l = L$. This is the *quantitative* manifestation of *self-similarity*, and it results from the absence of any other length scale, like ξ. In a way, this 'proves' the fractal behaviour at p_c.

If we are not at p_c, then we have a finite correlation length ξ, which becomes shorter by a factor b after renormalization. At the same time, the effective p' moves further away from p_c. Iterating the recursion relations many times now yields

$$M(L, \xi) = b^{lD} M(L/b^l, \xi/b^l) \qquad (65)$$

If $L \ll \xi$, then we may continue iterating until $b^l = L$. The whole lattice then becomes a single point, and we have

$$M(L, \xi) = L^D M(1, \infty) \propto L^D \qquad (66)$$

If $L \gg \xi$, then we should stop iteration at $b^l = \xi$, when $\xi_{\text{eff}} \equiv \xi/b^l$ becomes unity. If $p > p_c$, this means that p_{eff} is very close to unity, and the system looks uniform. In this case,

$$M(L, \xi) = \xi^D M(L/\xi, 1) \propto \xi^D (L/\xi)^d \qquad p > p_c \qquad (67)$$

the last step arising since for uniform systems $M(x, 1) \propto x^d$. Equations (66) and (67) justify in a way our Eq. (54).

If $p < p_c$ then p_{eff} is very small, and $M(x, 1) \propto x^{D_a}$ as for animals

($D_a = 1/\rho'$ is the fractal dimension for large animal-like clusters at $p < p_c$; see also Eq. (52)). Thus,

$$M(L, \xi) = \xi^D M(L/\xi, 1) \propto \xi^D (L/\xi)^{D_a} \qquad p < p_c \qquad (68)$$

For arbitrary L/ξ, Eq. (65) leads to finite size scaling, Eq. (55). Note that all these scaling equations do not require the detailed knowledge of the recursion relation which gives p' in terms of p. All that is required is the behaviour of lengths and masses under length rescaling. Similar ideas were introduced into the theory of critical phenomena in the mid-1960s by Kadanoff, and led Wilson to the introduction of the detailed renormalization group. The details of $p'(p)$ are, however, necessary for obtaining amplitudes, e.g. in Eq. (66).

4.4. LARGE CELL AND MONTE CARLO RENORMALIZATION

As we saw, the small cell renormalization schemes are only approximate. The excellent agreement found from Eqs. (63) and (64) for both p_c and ν is rather exceptional. For other lattices or in other dimensions there are usually stronger deviations. Only if we let b go to infinity can we expect that the renormalization result will approach the exact value. In a square lattice with $b = 5$, renormalization of the 25 sites in a 5×5 cell has to deal with 2^{25} different configurations, not an easy task (Reynolds *et al.*, 1980). To go to larger systems we use a Monte Carlo simulation to deal with a representative sample of all possible configurations. This method will be described below.

Another way to look at large cells is to evaluate the behaviour of two-dimensional strips of width b and infinite length. When b is of the order of ten, one can still do that exactly; for larger b one again needs Monte Carlo simulation. The main disadvantage is that this method is mainly restricted to two dimensions if one wants to avoid Monte Carlo simulation. For in three dimensions, an infinite bar of cross-section $b \times b$ already contains b^2 sites in each plane, which is easily more than the ten to twenty sites which can still be handled exactly. On the other hand, for two dimensions it gives very accurate estimates. It was crucial to have these estimates in order to believe the supposedly exact critical exponent of two-dimensional percolation in Table 2, Section 2.7. A review of this method was given by Vannimenus and Nadal (1984). Since much of this work on strips was done in Paris, the method can be called striptease, though names like 'phenomenological renormalization', 'Nightingale renormalization', 'transfer matrix approach', or just finite-size scaling, look more scientific.

We now return to Monte Carlo renormalization. We simulate randomly occupied cells on the computer by our well known Monte Carlo method, and then renormalize them.

How do we make the renormalization? We check whether the now rather large cell percolates, that is whether it is spanned by a cluster connecting top and bottom. This, however, was just what we described in the finite-size scaling Section 4.1 to find p_{av} and the width Δ. Thus the computer simulation is the same as in ordinary Monte Carlo work, only the analysis may differ. Similarly, for thermal critical phenomena the Monte Carlo renormalization technique first requires an ordinary simulation of the system; only the analysis of the resulting configurations may be different from finite-size scaling.

For the large cell renormalization group of percolation we thus see that the renormalized cell occupation probability p' is nothing but the spanning propability $\Pi(p)$. The b-dependent fixed point p^*, defined through $p^* = \Pi(p^*)$, is thus the intersection of the full curve in Fig. 17 with the diagonal from lower left to upper right. Since this curve becomes steeper as b increases, this intersection p^* approches p_c, and Eq. (57) implies that $p^* - p_c = const \cdot b^{-1/\nu}$. The slope $\lambda = d\Pi/dp$ at p^* diverges as $b^{1/\nu}$ (see Sec. 4.2), and thus the difference $\Pi(p^*) - \Pi(p_c) = (p^* - p_c)\lambda + ...$ may have a finite limit even when b becomes infinite (Ziff 1992).

From Eq. (57) it follows that $\Pi(p_c)$ is b-independent, and therefore it is supposed to be *universal*, whereas $\Pi(p^*) = p^*$ approaches the nonuniversal p_c. Such universal quantities have the same value for all lattices with the same dimensionality and the same spanning rule and shape. Also $(p^* - p_c)/\Delta$ and $(p_{av} - p_c)/\Delta$ should be universal ratios independent of b, for large b.

How can we understand this universality? Consider Π as a function of ξ and L. Close to p_c the renormalization will yield

$$\Pi(\xi, L) = \Pi(\xi/b, L/b) \tag{69}$$

without any b-dependent prefactor. As explained before Eqs. (56) or (65), such a prefactor arises only for quantities which diverge or vanish at p_c. As $p \to p_c$ and $L \to \infty$, the function $\Pi(\xi, L)$ approaches $\Pi(\infty, \infty)$. In that limit, rescaling by b does not change Π any more, since ∞/b is still infinite. Since $\Pi(\infty, \infty)$ does not depend on b, it should be determined by the fixed point, and therefore have the same universal value for all lattices with the same shape, spanning rule and dimension. A simple generalization of this argument shows that the whole function $\Phi(x)$ of Eq. (57) is a universal function of its argument $x = (p - p_c)L^{1/\nu}$, apart from a single scale factor for x. This explains the universality of ratios like $(p_{av} - p_c)/\Delta$, in which the scale factor cancels (see also p. 43).

Having stated that large cell renormalization group is for percolation nothing but a version of finite-size scaling, what have we learnt *new* from the former? Apart from giving us a conceptual framework to think about scaling, advanced renomalization group theory gave exact (as opposed to numerical) statements about universality of exponents, scaling functions and amplitude ratios, and also led to the analytic tool of the ϵ-expansion mentioned in Appendix B.

Returning now to Eq. (61) for $y = 1/\nu$ and $\Delta \propto b^{-1/\nu} = 1/\lambda$, we get

$$y(b) = \frac{\ln(1/\Delta)}{\ln b} - \frac{\text{Const}}{\ln b}$$

where ln denotes the natural logarithm and Const contains the proportionality factor. This result gives us, for every cell size b, an exponent $y = y(b)$. For $b \to \infty$ this effective exponent $y(b)$ should approach the true exponent $1/\nu$, since renormalization works exactly only for very large cells. Indeed, since Δ varies as $b^{-1/\nu}$ asymptotically, the above expression is compatible with this requirement; it shows us moreover that the effective $y(b)$ for finite cells differs from the asymptotic exponent $1/\nu = y(\infty)$ by, among other terms, a term proportional to $1/\log b$.

How can we in practice determine the asymptotic exponent $1/\nu$ from our widths $\Delta(b)$ for finite cell sizes b? We plot the ratio $\log(1/\Delta)/\log b$ versus $1/\log b$ and look for a smooth curve fitting these data and having a finite slope for $1/\log b$ going to zero. Instead of Δ one may also use $(2\pi)^{1/2} \Delta$ as argument of the logarithm to follow more closely the above derivation. In some examples the ratios $\log(1/\Delta)/\log b$ are larger than the extrapolated value $1/\nu$, in some cases they are smaller, but usually they can be fitted reasonably well on smooth curves with a finite slope at the intercept. Figure 23 gives an example for the triangular lattice.

What does that fitting procedure lead us to? If we look only at large b where the data are fitted reasonably well by the tangent on the curve with slope C and intercept $1/\nu$, we simply have

$$\frac{\ln(1/\Delta)}{\ln b} = \frac{1}{\nu} + \frac{C}{\ln b}$$

or

$$\Delta = \exp(-C)b^{-1/\nu}$$

This result, on the other hand, is nothing but our finite-size scaling result of Eq. (60b) for the widths of the threshold distributions. Thus, instead of going through the above analysis with $\log(1/\Delta)/\log b$ one may also simply take a log-log plot of width Δ versus cell size b or L and determine $y = 1/\nu$ from the asymptotic slope of that plot. (For fitting techniques see Appendices A.1–A.3.) In this sense, Monte Carlo renormalization of large cells is equivalent to finite-size scaling, if we look at these percolation problems. We have mentioned already that for thermal critical phenomena such simple relations do not hold in general. Thus again percolation works also as a particularly simple way to enter into modern phase transition research.

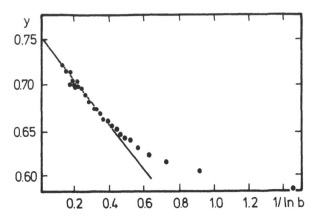

Fig. 23.   Results of large cell renormalization for the triangular lattice, using b up to 10 000 (see Eschbach *et al.*, 1981). The b-dependent effective exponents y, determined from the width of the distribution function for the threshold, are plotted versus $1/\ln b$ (solid circles). A tangent on the values for large b has the 'true' $y = 1/\nu$ for infinite systems as intercept. These data are compatible with the intercept being $0 \cdot 75$, corresponding to the supposedly exact $\nu = 4/3$.

4.5. CONNECTION TO GEOMETRY

Consider Fig. 19(c). It shows seven spanning clusters which contribute to the connectivity between the two edges of the 2×2 cell. In some sense, the first two of these (with probabilities p^5 and $p^4(1 - p)$) are more 'strongly' connected than the others: they will still connect even if we remove any one of the bonds in them. This is not true for the other configurations, all of which have *singly connected* bonds such that when one is removed the connectivity is destroyed. Such singly connected bonds also arise in larger cells. Since they are crucial for the connectivity, one might expect that they determine the exponent ν. This was indeed shown by Coniglio (1981) to be the case.

To quantify the removal of singly connected bonds from configurations like Fig. 19(c), Coniglio associated with each occupied bond an additional parameter π, which is the probability of not removing this bond from the system. We now consider the percolation problem as function of the two parameters, p and π. At p^*, we expect the spanning cluster to break into finite clusters for all $\pi < 1$. Thus, $\pi = 1$ is the fixed-point value of π when $p = p^*$. Consider now the recursion relation for $(1 - \pi)$, and expand the renormalized

$(1 - \pi')$ in powers of $(1 - \pi)$:

$$1 - \pi' = \Lambda(1 - \pi) + \Lambda_2(1 - \pi)^2 + \cdots \tag{70}$$

Since $(1 - \pi')$ is the probability of disconnecting the two edges of the cell, the linear term will arise only due to configurations which have singly connected bonds, and the coefficient Λ is exactly equal to their average number, denoted by $M_{sc}(b)$. Since varying π away from 1 is the same as varying p away from p^*, we also expect that $\Lambda = (d\pi'/d\pi)_{\pi^*} = (dp'/dp)_{p^*} = b^{1/\nu}$, and thus

$$M_{sc}(b) = b^{1/\nu} \tag{71}$$

Returning to Fig. 19, we see that indeed

$$p' M_{sc}(b) = 1 \times 4p^4(1 - p) + 3 \times 2p^3(1 - p)^2 + 2 \times [6p^3(1 - p)^2 + 2p^2(1 - p)^3]$$

and at $p = p^* = \frac{1}{2} M_{sc}(b) = \frac{13}{8}$, in agreement with Eq. (71). (Compare with results following Eq. (64).)

Equation (71) is a very important result. It shows that on average, the spanning cluster at p_c has *singly connected bonds on all length scales*. This will be an important ingredient in models for this cluster, discussed in the next chapter.

The procedure used here is applicable for many physical situations in which some additional problem is superimposed on top of the percolating lattice at p_c. In many cases, the physics is dominated by the singly connected bonds. Examples are given in Sections 5.6 and 7.7.

FURTHER READING

Bernasconi, J., *Phys. Rev. B*, **18**, 2185 (1978).
Burkhardt, T.W. and van Leeuwen, J.M.J., *Real Space Renormalization* (Heidelberg: Springer Verlag, 1982).
Coniglio, A., *Phys. Rev. Lett.*, **46**, 250 (1981).
Eschbach, P.D., Stauffer, D. and Herrmann, H.J., *Phys. Rev. B*, **23**, 422 (1981).
Fisher, M.E., in: *Proceedings of the Intern. Summer School Enrico Fermi, Course 51 Critical Phenomena*, Varenna, Italy (New York: Academic Press, 1971).
Harris, A.B., Lubensky, T.C., Holcomb, W.K. and Dasgupta, C., *Phys. Rev. Letters*, **35**, 327 (1975).
Kadanoff, L.P. *et al.*, *Rev. Mod. Phys.*, **39**, 395 (1967).
Levinshtein, M. E., Shklovskii, B.I., Sur, M.S. and Efros, A.L., *Zh. Eksp. Teor. Fiz.*, **69**, 386 (1975); English translation: *Soviet Phys. JETP*, **42**, 197 (1976).
Nakanishi, H. and Reynolds, P.J., *Phys. Lett. A*, **71**, 252 (1979).

Reynolds, P.J., Stanley, H.E. and Klein, W., *Phys. Rev. B*, **21**, 1223 (1980)
Reynolds, P.J., Stanley, H.E. and Klein, W., *J. Phys. C*, **10**, L 167 (1977).
Vannimenus, J. and Nadal, J.P., *Phys. Reports*, **103**, 47 (1984).
Wilson, K.G., *Phys. Rev. B*, **4**, 3174 and 3184 (1971).
Ziff R.M., *Phys. Rev. Letters*, **69**, 2670 (1992).

CHAPTER 5

Conductivity and Related Properties

So far, we have concentrated only on *statistical* and *geometrical* properties of percolation clusters. As we indicated in the introduction, many applications of the theory concern other properties, which arise by superimposing some additional processes which occur on top of the dilute sample. Examples included the propagation of forest fires, the flow of fluid in a porous medium and the diffusion of particles through a dilute system. The remaining chapters of this book aim to review some of these applications more quantitatively.

In this chapter we discuss various problems associated with the conductivity of dilute systems, relevant for example to the fluid conductivity (sometimes called permeability) of the porous medium discussed in Section 1.3, or the electrical conductivity of metal–insulator alloys. The conductivity-related properties of *random resistor networks* turn out to open a whole Pandora's box of new geometrical properties, which yield a multitude of new fractal dimensions and lead to *multifractality*. After describing these in general terms, we present *recursive fractal models*, which imitate the percolation clusters, and explain how the small cell renormalization group in Section 4.2 can be used to calculate some of the transport exponents. We end with brief reviews of continuum percolation and of elastic properties.

5.1. CONDUCTIVITY OF RANDOM RESISTOR NETWORKS

Let us go back to the squares of Fig. 1, which are randomly occupied or empty. We regard every occupied square as a piece of copper, whereas every empty square is regarded as insulating. An electric d.c. current can only flow between copper squares having one side in common, not between occupied squares touching at a corner only or separated by even longer distances. How much current flows through the lattice if a unit voltage is applied between the top line and the bottom line of the lattice (see Fig. 24)? We call this current due to a unit voltage the *conductance* of the sample.

We take the whole lattice to have a rectangular shape of $L \times N$ squares, with N being the length of the top (and bottom) row to which a uniform

Fig. 24. Definition of the conductance of a random conductor network. All copper squares in the topmost row of the lattice are connected to a heavy copper bar (no loss of energy in the bar), and so are all squares in the bottom row. A battery then applies a unit voltage between these two bars. The resulting electrical current is called the conductance.

voltage is applied (see Fig. 24). Both N and L are very large. If the system is uniform and homogeneous like a sheet of copper, then the conductance is proportional to N and inversely proportional to L. In d dimensions it is still inversely proportional to L, but it is now proportional to the cross-section, N^{d-1}, of the sample. Thus the conductance is proportional, in d dimensions, to N^{d-1}/L, and the factor of proportionality is called the *conductivity* Σ of the material, which is now independent of size and shape. For a square or cubic shape, one has $L = N$ and thus the conductivity Σ is L^{2-d} multiplied by the current produced by a unit voltage. (We set the distance between neighbouring points on the lattice equal to unity.)

It is obvious that for large lattices we have zero conductivity if no infinite network of neighbours is present, that is for $p < p_c$. When p is appreciably larger than p_c, nearly all copper squares have clustered together to form one infinite network, and the conductivity Σ as well as the fraction P of sites in the infinite network increase roughly linearly with concentration p. At $p = 1$ we have, of course, $P = 1$, and Σ reaches the conductivity of bulk copper. If we set this copper conductivity equal to unity, then $\Sigma(p = 1) = P(p = 1) = 1$. In other words, our copper squares carry a unit current if a unit voltage is applied to two opposing sides.

Because of this close relationship between conductivity $\Sigma(p)$ and mass $P(p)$ of the infinite network it would be nice if they were proportional (and because of our normalizations that would mean identical) over the whole range of p. The first experiment, by Last and Thouless (1971), showed that this is not the case. They measured the current through sheets of graphite paper with randomly punched holes. Their results, shown schematically in Fig. 25 indicate clearly that the two quantities Σ and P are not proportional. The conductivity versus concentration curve seems to end at the threshold with zero slope although a plot of P versus p has infinite slope there.

Why is this so? Figure 26 shows a section of a bond cluster on the square lattice at p_c. If each bond on this cluster represents a resistor, and if we put a voltage between the upper and lower ends of the cluster, then one identifies many bonds which will carry no current, because they lead nowhere. These bonds (denoted by thin lines in Fig. 26) are called 'dangling', or 'dead ends'.

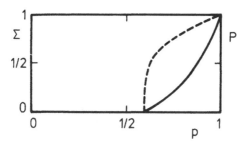

Fig. 25. Conductivity Σ of conducting paper with holes randomly punched in it (solid line; from Last and Thouless, 1971). The dashed line gives the percolation probability P. Obviously, the two quantities vanish with different exponents μ and β, though at the same threshold p_c.

Fig. 26. Section of cluster at p_c (after Stanley, 1977). Thin lines represent 'dangling' bonds; thick lines represent singly connected bonds; dotted lines show bonds on 'blobs'.

When we erase all the dangling bonds, we are left with the 'backbone'. Every internal bond on the backbone has at least two independent routes that lead from it to the edges of the cluster. Except for rare symmetric situations, all the backbone bonds will carry some current when we put a voltage between the upper and lower edges of the cluster. As we shall show in the next two sections, most of the mass of the infinite network at the threshold belongs to dead ends, not to the backbone. Thus most of the mass contained in P makes no contribution to the conductivity Σ, and therefore the critical exponent for Σ differs from the β for P.

We denote the conductivity exponent by μ:

$$\Sigma \propto (p - p_c)^{\mu} \qquad (72)$$

for $p \rightarrow p_c$. Often this exponent is also called t but here we need t as the symbol for time (and earlier we used it for the perimeter.) Perhaps we should leave it as an exercise to the reader to derive a simple and exact scaling relation between this new exponent μ and the old ones like β, ν, etc., for the people working in this field have not, at present, agreed on such a relationship. Various conjectures for such relationships have been suggested over the last few years, but none of them seems to be exactly true. We shall mention one of these, by Alexander and Orbach, in Section 6.2. In the absence of exact relations, we must at present consider μ as a new independent basic exponent.

How does the conductivity Σ depend on the size of the sample? From Eq. (56), we expect finite size scaling to apply in the form

$$\Sigma(L, \xi) = \xi^{-\mu/\nu} S(L/\xi) \propto \begin{cases} \xi^{-\mu/\nu} & L \gg \xi \\ L^{-\mu/\nu} & L \ll \xi \end{cases} \tag{73a}$$

Thus, the statements made above that Σ is independent of the size of the sample and can be described by the law (72) are really true only for sufficiently large samples, for which $L \gg \xi$. For smaller sizes, or at p_c, we expect the size-dependent behaviour

$$\Sigma \propto L^{-\mu/\nu} \tag{73b}$$

Figure 27 shows some conductivities Σ as functions of system size L right at the percolation threshold. Indeed, these Monte Carlo simulations confirm the power-law dependence of Σ on L, and yield μ/ν near $0 \cdot 975$ in $d = 2$ and of order $2 \cdot 3$ in $d = 3$.

How does one conduct such Monte Carlo studies? The easy part is to produce a lattice with conducting and insulating sites or bonds. Such a lattice is called a 'random resistor network'. The difficulty lies in calculating the con-

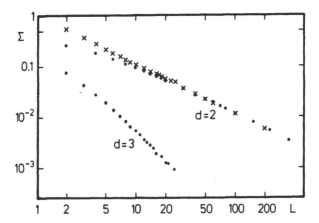

Fig. 27. Variation of the conductivity Σ with system size L in three and two dimensions right at the percolation threshold. The two data sets for two dimensions refer to different geometries.

ductance of that lattice. Kirchhoff's rules tell us that for every loop of conductors the sum of the voltages is zero, and that for every site the sum of the currents flowing into it is also zero. The resulting system of coupled linear equations for the voltages at every site can be written as

$$I_i = \sum_j (V_j - V_i)\sigma_{ij} \qquad (74)$$

where V_i is the voltage at site i, I_i is the external current going into the site i (it is non-zero only at the external terminals, and zero on all other sites) and σ_{ij} is the conductance of the bond connecting the nearest neighbour sites i and j. For the simple dilute resistor network, $\sigma_{ij} = 1$ for an occupied bond and zero otherwise. This system can be solved by relaxation methods (Kirkpatrick 1973). The closer we are to the percolation threshold, the more iterations we need in this relaxation before we get the solution with the desired accuracy. While this 'critical slowing down' is of interest in itself, it also increases the computer time. That difficulty can be avoided if in two dimensions special transformations are used (see, e.g., Gingold and Lobb, 1990) which allow the conductivity to be calculated by going just once through the lattice. In a similar spirit, but with an algorithm which is also useful for more than two dimensions, Derrida *et al.* (1984) calculated the conductivity of narrow strips (and bars) in their combination of Monte Carlo simulation with the 'striptease' renormalization of Section 4.4. We refer to these papers for more computational details.

A special-purpose computer has been built by Normand *et al.* (1988) which can calculate only conductivities of such strips, but with Cray-computer speed. In two dimensions, its result $\mu/\nu = 0\cdot9745 \pm 0\cdot0015$ was found by extrapolating conductivities $\Sigma \propto L^{-\mu/\nu}$ in strips of width L and confirmed with better accuracy older Monte Carlo estimates. (There are still some discrepancies with extrapolations from exact series expansions; see Adler *et al.*, 1990.)

Instead of working with resistors and insulators, one can also study a mixture of resistors and superconductors. In that problem each site has either zero resistivity (infinite conductivity) with probability p, or has a probability $1 - p$ of being a normal resistor. All quantum-mechanical aspects of real superconducting materials are neglected, of course; we still deal only with geometry. The situation is formally similar to that of random resistor networks except that zero conductivity is replaced by zero resistivity. The conductivity of this network of superconductors and normal resistors is infinite whenever an infinite network of superconducting sites is formed, that is for $p > p_c$. It is finite below p_c; thus it is possible that this conductivity diverges as p_c is approached from below. In two dimensions it diverges with the same exponent μ with which the conductivity of resistor–insulator networks vanishes. In general, this nice duality rule is not valid. For example in three dimensions, where roughly $\mu = 2$, the superconducting exponent, often denoted as s, is about $0\cdot73$.

One can also mix superconductors with normal resistors. More generally, scientists have mixed resistors of different strength and derived scaling laws for that situation. Also, resistors can be replaced by diodes or by capacitors to model dielectric materials. In an electric field, real materials are electrically polarized, i.e. one end of the sample is positively charged, the other negatively. This effect leads to a polarization of the metallic clusters in the dielectric, and to a divergence of the dielectric constant as these clusters approach percolation. If in addition to an electric field we apply a magnetic field, the conduction electrons are forced into a direction perpendicular to the current. This 'Hall' effect has also been studied in disordered networks. We refer the reader to current literature for more details.

5.2. INTERNAL STRUCTURE OF THE INFINITE CLUSTER

As discussed above, the infinite cluster may be divided into boxes of size ξ. Inside each box, the geometry of the cluster resembles that of the infinite cluster at p_c. That cluster is very 'weak': removal of a few bonds can break it into finite clusters, and disconnect the edges of the system from each other (see Section 4.5). This weakness of the cluster led Skal and Shklovskii (1975) and de Gennes (1976) to postulate that within each box of size ξ there is practically only one chain of bonds that connects its opposite edges. This led to a model which has a network of *nodes*, at distance ξ from each other, connected by effectively one-dimensional *links*. Although this picture turns out to be correct for dimensions $d > 6$, when the probability of finding large loops is very small (see Appendix B), it is too simplified for $d < 6$. Figure 26 shows a section of a cluster at p_c. This section should be representative of the 'links' on scales smaller than ξ. Clearly, this cluster does not seem to be a one-dimensional channel of occupied sites, but resembles more a network of roads, with many parking places and dead ends.

As noted, only the backbone matters when we consider the conductivity of the cluster. Within the backbone, we identify two types of bonds. As we showed in Section 4.5, some bonds are 'singly connected'. These bonds are drawn with thick lines in Fig. 26. Whenever we 'drive' from top to bottom, we must go through these 'bottle necks'. In terms of the resistor network, these bonds carry the full current which goes through the circuit.

If all the bonds on the backbone were singly connected, then we would recover the simple 'links and nodes' model. However, it is clear from Fig. 26 that the singly connected segments are often separated by structures which contain several routes in parallel, or *loops* (these may be visualized as the streets of villages or towns on our route from top to bottom). Such *multi-connected* pieces between *singly*-connected parts are called 'blobs', and are represented by dotted lines in Fig. 26.

As we showed in Section 4.5 (Eq. (71)), the number of singly connected bonds on the backbone, between two edges of a box of size $L < \xi$, is not zero,

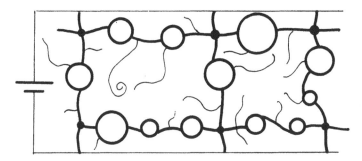

Fig. 28. Schematic picture for the links (one-dimensional chains), nodes (crossing points of the links) and blobs (dense regions with more than one connection between two points; shown as circular here) of the infinite cluster slightly above the threshold. The distance between the nodes as well as the maximum blob diameter are assumed to be of the order of the correlation length. The thin lines are the dead ends, for clarity only very few of them are shown. Most of the material is in the dead ends, the rest is called the backbone. Most of the backbone mass belongs to blobs.

but rather proportional to $L^{1/\nu}$. For a box of size ξ, this number grows as $\xi^{1/\nu} \sim (p - p_c)^{-1}$. This exact result was consistent with numerical simulations, e.g. by Pike and Stanley (1981). The fact that there exist singly connected bonds between edges at distance ξ confirms the picture which allows only one main route, or generalized link, between neighbouring nodes. However, the link between nodes is now more complicated, having both singly connected bonds and blobs. Although the length of each singly connected segment is finite (of order 2–3 in two dimensions), the blobs have all possible sizes up to ξ.

All the above arguments led to the 'links–nodes–blobs' picture (Stanley, 1977). This picture is summarized schematically in Fig. 28: nodes at distance ξ are connected by generalized links, which contain both singly connected bonds and blobs. All of these 'main roads' also have exits to dead ends. As we shall see below, this geometrical picture turns out to be very useful in discussing physical properties of the percolating system.

5.3. MULTITUDE OF FRACTAL DIMENSIONS ON THE INCIPIENT INFINITE CLUSTER

We now restrict ourselves to scales $L \ll \xi$, or to $p = p_c$. We have already identified several *subsets of bonds* (or sites) on the incipient infinite cluster. Since at p_c there is no basic length scale, we expect the 'masses' of each of these subsets also to scale as a power of L. Indeed, measurements (e.g. with frames of size $L \times L$ on pictures like Fig. 2) of the backbone mass yield results of the form

$$M_B(L) \propto L^{D_B} \tag{75}$$

For all $d > 1$ one finds that $D_B < D < d$. For example, $D_B \simeq 1.74$ in $d = 3$. For $d > 6$ one has $D_B = 2$, resulting from the fact that at high dimensions there are very few large loops, and the links behave like random walks which practically do not intersect themselves, or 'self-avoiding walks'. The fact that $D_B < D$ means that the ratio M_B/M approaches zero for large L, implying that practically all the mass of the cluster belongs to the dangling ends. This fact might be crucial for secondary oil recovery, if one happens to be close to the percolation threshold mentioned in Section 1.3. In secondary oil recovery, a common technique is to drill a second well some distance L away from the production well, and push water into this new well. The water is expected to push the oil out of the production well. Clearly, this may work only if both wells are on the same cluster. If the oil in the dangling bonds is incompressible, then this procedure will never push it out. At best, only the oil in the backbone will be pushed. Thus, the estimate with L^D used in Section 1.3 is much too optimistic: at most only a fraction $M_B/L^d \sim L^{D_B - d}$ will come out, and $d - D_B \simeq 0.4$ or 1.2 for $d = 2$ or 3!

In reality, the situation is even worse: the front separating the water and the oil is unstable, and the water reaches the end of a blob without clearing all the oil in it. This problem, of 'viscous fingering' on percolation clusters, has recently been studied by both simulations and model experiments (Oxaal *et al.*, 1987).

We now return to the quantitative characterization of subsets on the cluster. We imagine again this cluster to be a random resistor network. If we want to know which bonds on the cluster carry the full current, we need to count only the singly connected bonds. This is relevant if each resistor is an electrical fuse, which may burn if a large current goes through it. Clearly, the singly connected bonds will be the first to burn. This is why Stanley (1977) called them 'red bonds'. From Eq. (71), the number of singly connected bonds behaves as

$$M_{SC}(L) \propto L^{D_{sc}} \qquad (76)$$

with $D_{SC} = 1/\nu$. Clearly, $D_{SC} < D_B$ for $1 < d < 6$, indicating that practically all the mass of the backbone resides in the blobs.

Another property of the cluster concerns the *resistance* between the two edges, or between the two terminals on top and bottom of Fig. 24 (or Fig. 26). Since $L \ll \xi$, the resistance $1/G$ is also a power of L,

$$\frac{1}{G(L)} \propto L^{\tilde{\zeta}_R} \qquad (77)$$

As we demonstrate below, the exponent $\tilde{\zeta}_R$ is directly related to the exponent μ/ν via the relation

$$\frac{\mu}{\nu} = d - 2 + \tilde{\zeta}_R \qquad (78)$$

Since this relation contains d, it is also 'hyperscaling', and it is valid only for $d \leqslant 6$.

Before we end this list, we mention another family of substructures on the cluster. This concerns *linear polymers*, or *self-avoiding walks*. A linear polymer is constructed as a flexible chain of monomers, which never intersects itself. The way to simulate such chains on a lattice is similar to that described in Section 1.4 for random walks, or diffusing particles: whenever the walker reaches a lattice point, it may proceed with equal probability to the neighbouring sites. However, unlike the diffusion case, the polymer walker is very clever, and it never walks back. In addition, it never steps on a site which it has already visited, since that site is already occupied by a monomer of the growing polymer. Imagine now such a polymer, going through our cluster. Clearly, it cannot go into any of the dangling bonds, since it is not allowed to step on itself on the way out. It is thus constrained to move on the backbone. Also, it must go through all the singly connected bonds. However, inside each blob there are several routes through which the polymer may go. We thus end up with a *distribution of lengths* of the possible self-avoiding walks between the two end points. The shortest of these represents the *minimal path* l_{min} (often also called the *chemical distance*; Havlin, 1984) needed for going on the cluster from one end to the other. Herrmann and Stanley (1988) used 10^4 computer hours to measure l_{min} at p_c in two and three dimensions. As expected, this minimal path also has a fractal behaviour,

$$l_{min}(L) \propto L^{D_{min}} \qquad (79)$$

with $D_{min} \approx 1 \cdot 13$, $1 \cdot 34$ in $d = 2, 3$.

We shall describe below several physical applications of the minimal path. One obvious application concerns the forest fires mentioned in Chapter 1: if it takes a unit time to transfer the fire from one tree to its neighbour, then the minimal time for the fire to spread to the other end scales as $l_{min}(L)$.

In addition, one can study the *maximal self-avoiding walk*, which scales as $l_{max}(L) \propto L^{D_{max}}$, and the *average self-avoiding walk*, defined as $l_{SAW} \propto L^{D_{SAW}}$. The latter exponent is sometimes denoted as $D_{SAW} = 1/\nu_{SAW}$.

One can easily convince oneself of the following hierarchy of all these fractal dimensionalities, or exponents:

$$d - \beta/\nu = D \geqslant D_B \geqslant D_{max} \geqslant D_{SAW} \geqslant D_{min} \geqslant \tilde{\zeta}_R \geqslant D_{SC} = 1/\nu \qquad (80)$$

For example, l_{max} may have fewer bonds than the backbone, since the walk is not allowed to cross itself. Similarly, $D_{min} \geqslant \tilde{\zeta}_R$ since l_{min} represent only one path through each blob, and the resistance $1/G$ contains additional bonds in parallel, which may lower the net resistance. The resistance of the singly connected bonds, equal to M_{SC}, is clearly less than $1/G$ since $1/G$ contains the additional resistance of the blobs, hence $\tilde{\zeta}_R \geqslant D_{SC}$. Table 2 contains a summary of known values for all these exponents.

All the discussion so far has been restricted to $L \ll \xi$. If $L \gg \xi$, then for $p > p_c$ we saw that we can divide the system into $(L/\xi)^d$ boxes, which may be treated homogeneously. Thus, Eq. (55) shows that the total mass obeys a scaling form in terms of the variable L/ξ. A similar form is expected to hold for all other properties, e.g.

$$M_B(L) = L^{D_B} M_B(L/\xi) \tag{81}$$

which varies as L^{D_B} for $L \ll \xi$ and as $\xi^{D_B - d} L^d$ for $L \gg \xi$.

The L-dependence for $L \gg \xi$ is not always like L^d. For example, we expect $l_{\min}(L)$ to depend *linearly* on L for a homogeneous system, hence

$$l_{\min}(L) = L^{D_{\min}} \lambda_{\min}(L/\xi) \tag{82}$$

and $l_{\min}$ varies as $L^{D_{\min}}$ for $L \ll \xi$ and as $\xi^{D_{\min} - 1} L$ for $L \gg \xi$. The situation for $M_{SC}(L)$ is different: by Eq. (76), $M_{SC} \propto L^{1/\nu}$ for $L \ll \xi$. For $L \gg \xi$ and $p > p_c$ we expect no singly connected bonds, and M_{SC} approaches zero.

The situation is somewhat more complicated for the resistance. As mentioned, in a homogeneous material of size L^d, the resistance between two parallel planar electrodes on two parallel edges grows linearly with the distance L between them and decreases as the inverse of the cross section, L^{d-1}. Thus, the conductance ($= 1/$resistance) of the system made of $(L/\xi)^d$ boxes will behave as

$$G \propto \xi^{-\tilde{\zeta}_R} (L/\xi)^{d-2} \propto L^{d-2} \xi^{-(d-2+\tilde{\zeta}_R)} \tag{83}$$

The conductivity, defined as

$$\Sigma \equiv L^{2-d} G$$

therefore behaves as $\Sigma \propto \xi^{-(d-2+\tilde{\zeta}_R)} \propto \xi^{-\mu/\nu}$, thus proving Eq. (78). From the values of μ/ν quoted above, we find $\tilde{\zeta}_R \simeq 0 \cdot 975$ and $1 \cdot 3$ in $d = 2$ and 3.

For $L \gg \xi$ and $p < p_c$, we expect crossover to the behaviour of *lattice animals*. It turns out that large loops are very rare in animals. Therefore, the backbone is well described by a tenuous singly connected link, and one has $D_B = D_{\max} = D_{SAW} = D_{\min} = \tilde{\zeta}_R = D_{SC} \simeq 1 \cdot 17$ and $1 \cdot 36$ in $d = 2$ and 3 (Havlin *et al.*, 1984).

All the above was valid for $d \leqslant 6$. What happens for $d > 6$? At high dimension, we argued in Section 2.1 that the tree-like structures dominate, and loops are relatively unimportant. For such structures, the backbone is practically the same as a random walk (each step goes along a new axis), with a fractal dimension equal to $D_B = 2$ (see Section 1.4). Since loops are not important, we have

$$D_B = D_{\max} = D_{SAW} = D_{\min} = \tilde{\zeta}_R = D_{SC} = 2$$

Using $D_{SC} = 1/\nu$ (Eq. (76)), this geometrical argument yields $\nu = 1/2$ for these dimensions. We can then use Eq. (49) to find that for such high dimensions $D = 1/(\sigma\nu) = 4$. In Appendix B we show that these fractal dimensions remain self-consistent for all $d > 6$, since the density of points at which loops are formed decreases as the size of the loops grows.

We thus conclude that the hyperscaling Eqs. (53) and (54) are not valid for $d > 6$. Instead, we have $M \propto L^4$ for $L \ll \xi$ and $M \propto (p - p_c) L^d \propto \xi^{-2} L^d$ for $L \gg \xi$. In the latter case, the mass within a box of size ξ is proportional to

$\xi^{d-2} = \xi^{d-6}\xi^4$, and the extra factor ξ^{d-6} has to do with the number of different infinite networks in that box (see Appendix B.1).

Similarly, for $d > 6$ Eq. (78) is replaced by the Bethe lattice result $\mu/\nu = 6$, and $\mu = 3$. The resistance $1/G$ is proportional to $L^{\tilde{\zeta}_R} = L^2$ for $L \ll \xi$, and to $L^{2-d}\xi^6 = (L/\xi)^{2-d}\xi^{6-d}\xi^2$ for $L \gg \xi$ (Aharony *et al.*, 1984).

5.4. MULTIFRACTALS

Whenever one deals with random systems, one only has statistical information about their properties. This information is usually expressed via distribution functions. For example, $n_s(p)$ denotes the *average* number (per site) of clusters containing s sites each (see Section 2.4). This distribution function allows us to calculate average cluster properties. In particular, it allows us to derive the *average moments* of the cluster size distribution, M_k. As Eq. (31) indicates, these moments have simple scaling behaviour. Using Eq. (53), Eq. (31) may be rewritten as

$$M_k \propto |p - p_c|^{(d-kD)\nu} \propto \xi^{kD-d} \qquad (84a)$$

The exponent on the righthand side is linear in k, implying a constant gap between exponents of consecutive ks. This constant gap indicates that all the moments are determined by a single basic mass, $s_\xi \propto \xi^D$, which scales with the single fractal dimension, D. This single gap scaling is also a direct result of the fact that $n_s(p)$ can be written in a scaling form like Eq. (33), with a single scaling variable.

In fact, most of the power-law expressions discussed in this book so far concern *average* properties. For example, the expression $M(L) \propto L^D$ for the mass of the infinite incipient cluster at p_c within a box of linear size L (Eq. (54), Section 1.3 and Fig. 16) is obtained after averaging over many such boxes. The values for individual boxes will fluctuate around the average, and the complete information is contained in their *distribution function*. Fortunately, this distribution function also turns out to be simple, with a constant gap between the exponents for its average moments. Furthermore, one can also quantify the above fluctuations, by looking at *cumulants* of the distribution. The second cumulant is simply the mean square deviation from the average, $\Delta M^2 = \langle (M - \langle M \rangle)^2 \rangle$. This is a measure of the mass fluctuations, reflected by some boxes having more mass and some others having less mass (more 'lakes'). ΔM^2 scales as L^{2D}, and all the cumulants are also determined by the single exponent D. The ratio $\Delta M^2/M^2$, which is sometimes called 'lacunarity', is size independent (for large L), and serves as a measure for the deviation of the fractal from being homogeneous.

Another example involves the resistance between two terminals at a distance L apart. In Eq. (77) we made a statement about the *average* of this resistance, implying that one should prepare many samples of clusters at p_c, or many sections of the same cluster (by moving the two terminals around), and average over the measured resistance of all of them. The results of these measurements can be collected into a histogram, giving the number of samples which gave a resistance in each range, and this yields the *resistance distribution* $P(1/G, L)$. It turns out that, like $n_s(p)$, this distribution also obeys a single variable scaling, i.e. all the moments of the two terminal resistance have a constant gap,

$$\langle (1/G)^k \rangle \propto L^{k\bar{\zeta}_R} \tag{84b}$$

Similar laws apply for many of the other quantities described in the last section.

The situation changes when instead of averaging over all possible configurations we concentrate on a single configuration. In this case the distributions of properties on the cluster may become more complicated. An illuminating example concerns the *distribution of currents* on the bonds of the random resistor network. Imagine again that we connected the two ends of Fig. 26 to a unit voltage. Denote now the current through the bond b by I_b, and the total current by I. These currents will clearly have a complicated distribution, with I_b varying from the full I (on singly connected bonds) to very small values (on some remote blob bonds). The currents on the dangling bonds are zero. Consider now the moments

$$M_q(L) = \sum_b (I_b/I)^{2q} \tag{85}$$

(The experts should note that only the normalized ratios M_q/M_0 are moments; since the denominator only represents a trivial shift in the exponents, we keep the loosely defined name 'moment' for M_q). It turns out that (at least for $q > 0$) $M_q(L)$ scales as a power of L,

$$M_q(L) \propto L^{y(q)} \tag{86}$$

However, the family of exponents $y(q)$ does not have a constant gap, and we are not aware of any linear relation among the different $y(q)$. One thus needs an infinite number of independent exponents to characterize the current distribution. The distribution of currents is therefore called *multifractal*. In the rest of this section we discuss this distribution in some detail. Since this discussion is somewhat complicated, and is not required for the rest of the book, the less interested reader may now move to the next section.

It is easy to see that different moments scale differently with L. When $q \to \infty$, all the terms with $I_b < I$ (arising from blobs) will drop out (a number less than one becomes very small when raised to a very large power). Thus, in this limit the sum will contain only the contributions from the singly connected bonds, for which $(I_b/I)^{2q} \equiv 1$, since each such bond carries the full current. Thus,

$$M_\infty(L) \equiv M_{SC}(L) \propto L^{D_{SC}}$$

and $y(\infty) = D_{SC}$. When $q \to 0$, every term in the sum which arises from $I_b \neq 0$ will contribute 1; hence (neglecting the few non-current-carrying bonds on the backbone)

$$M_0(L) = M_B(L) \propto L^{D_B}$$

and $y(0) = D_B$. Since $(I_b/I)^{2q}$ is a monotonically decreasing function of q, it follows for $q > 0$ that

$$M_{SC}(L) \leqslant M_q(L) \leqslant M_B(L)$$

or that

$$\frac{\ln M_{SC}}{\ln L} \leqslant \frac{\ln M_q}{\ln L} \leqslant \frac{\ln M_B}{\ln L}$$

For $L \gg 1$, the two bounds reach the constant values D_{SC} and D_B. Therefore, $\ln M_q/\ln L$ must also reach some constant value, which we denote by $y(q)$ (the notations $\tilde{\psi}(q)$ and $-x_q$ also appear in the literature). This requires that M_q is also a power of L, and specifically yields Eq. (86). The exponent $y(q)$ is monotonically decreasing from D_B (at $q = 0$) to D_{SC} (at $q \to \infty$). It is thus *not linear* in q, and therefore the current distribution is not as simple as that of the cluster sizes. In fact, the moments $M_q(L)$ are not all determined by a single basic mass, as was the case in Eq. (84a). Instead, different moments are dominated by different subsets of the bonds. For $q \to \infty$, we saw that M_q is dominated by the singly connected bonds, which are a subset of the backbone bonds that dominated M_0. We know no simple relation between D_{SC} and D_B, and thus need more than one scaling variable to describe the current distribution. This is the origin of *multifractality*: every moment M_q is dominated by a different fractal subset of the backbone, with a different fractal dimension related to $y(q)$, and there is no linear relation among the various y's. Without blobs we would have $M_{SC} = M_B$ and thus $y(q) = D_B = D_{SC}$ for all q: multifractality would then be impossible (see Eq. (86)).

Two additional values of q are of special interest. If each bond has a resistance $1/\sigma_b$, then the total energy dissipation in the network (responsible, for example, for the Joule heating of resistors) is given by

$$P = \sum_b \frac{I_b^2}{\sigma_b} = \frac{I^2}{G}$$

where $1/G$ is the total resistance between the ends. If all the occupied bond resistances are the same, this implies that

$$\frac{1}{G} = \frac{M_1(L)}{\sigma_b} \propto L^{y(1)}$$

and hence $y(1) \equiv \tilde{\zeta}_R$.

If we allow the resistances $1/\sigma_b$ to *fluctuate* (e.g., due to thermal noise), then $1/G$ will also fluctuate. A theorem by Cohn, which we do not prove here, shows that the deviation of $1/G$ from its average, $\delta(1/G) \equiv 1/G - \langle 1/G \rangle$, is given by

$$\delta(1/G) = \sum_b \left(\frac{I_b}{I}\right)^2 \delta(1/\sigma_b)$$

If the mean square fluctuation of each resistor is $\langle (\delta(1/\sigma_b))^2 \rangle = \Delta$, and these fluctuations are uncorrelated to each other, then the mean square fluctuations (or the 'second cumulant') of the total resistance becomes

$$\langle [\delta(1/G)]^2 \rangle = \sum_b \left(\frac{I_b}{I}\right)^4 \Delta = \Delta M_2 \propto L^{y(2)}$$

Similarly, higher cumulants are given by M_q with higher q's. The fluctuations in $1/G$ are related to the fluctuations in I for a fixed voltage. Thus, M_2 is proportional to the *noise* in the network (Rammal *et al.*, 1985).

Note that the mean square deviation $\langle [\delta(1/G)]^2 \rangle$ represents the variation of $1/G$ for a *single realization* of the sample and of the two terminals, as a result of fluctuations in the *individual* σ_b's. This is not to be confused with the variations between *different* samples, leading to the 'unifractal' result Eq. (84b).

To study multifractality quantitatively, denote $i = I_b/I$, and let $n(i, L)$ be the number of bonds which have a current fraction i. Thus, Eq. (85) becomes

$$M_q(L) = \sum_i n(i, L)i^{2q}$$

with $0 < i \leqslant 1$. For finite q, $n(i, L)i^{2q}$ is very small for both $i \to 0$ and $i \to 1$, and has a peak at some intermediate value i_q. Assuming that this peak is large, we approximate M_q by

$$M_q \simeq n(i_q, L)i_q^{2q} \tag{87}$$

and i_q is given by the solution of

$$\left.\frac{\mathrm{d}\ln n(i,L)}{\mathrm{d}\ln i}\right|_{i=i_q} = -2q \tag{88}$$

At p_c (or for $L \ll \xi$), we are used to assume that every function of L must be a power law. If this is true, then we can try the power laws

$$i_q^2 \propto L^{-\alpha(q)} \qquad n(i_q, L) \propto L^{f(q)} \tag{89}$$

The new exponent $f(q)$ may be interpreted as the fractal dimension of the subset of bonds that dominates the moment M_q. Substituting in Eq. (87) we recover Eq. (86), with

$$y(q) = f(q) - q\alpha(q) \tag{90}$$

Equation (88) now reads

$$\frac{\mathrm{d}f}{\mathrm{d}\alpha} = q \tag{91}$$

These two equations may be converted into

$$f(q) = y(q) + q\alpha(q) \qquad \alpha(q) = -\frac{\mathrm{d}y}{\mathrm{d}q} \tag{92}$$

and represented as a single function $f(\alpha)$. If a distribution function is really multifractal (for large L), then the L-dependent function $n(i, L)$ can be replaced by the L-independent (and presumably universal) function $f(\alpha) = f(q(\alpha))$. For large L, f and α are respectively close to $f = \log n / \log L$ and $\alpha = -\log i^2 / \log L$. Thus, plots of f versus α for different L should fall on a single *data-collapsed* curve. Indeed, such data collapse has been observed for $q \geqslant 0$ in many computer simulations. $f(\alpha)$ increases from D_{SC} at $\alpha = \alpha_{min} = 0$ (corresponding to $q = \infty$) to D_B for $\alpha = \alpha_0$ (corresponding to $q = 0$).

The situation for $q < 0$ is much more complicated. When looking at $q < 0$, we must omit all the terms with $I_b = 0$ in Eq. (85). The remaining sum is dominated, for large negative q, by the *smallest* currents. It turns out that these small currents decay exponentially with the cluster size, and thus the power law (86) no longer holds (see Blumenfeld *et al.* (1987) for more details).

Another example of a multifractal distribution on a single realization of the backbone is that of the lengths of the self-avoiding walks (Furuberg *et al.*, 1987).

5.5. FRACTAL MODELS

As we have seen so far, it is not easy to obtain many exact analytical results for cluster properties. Most of our knowledge comes from complex numerical calculations. In many cases in natural science it is very useful to invent simple mathematical models, on which one can do analytical calculations. Such models try to capture some of the important features of the problem at hand, and hope that these features are sufficient so that the resulting model also predicts other features reasonably well.

In what we have seen so far, the geometry of the infinite cluster at p_c is fractal. We also saw that it is characterized by many fractal dimensions, which describe subsets of the cluster sites (or bonds) necessary for calculating different cluster properties. In what follows, we describe several families of *recursive geometrical fractal models* which were proposed to imitate some aspects of the fractal geometry of the infinite cluster. These models are fractal, but they are not random. The lack of randomness enables us to perform exact analytical calculations, even though the structures are fractal and not translationally invariant.

To demonstrate the construction of a recursive fractal, consider the *Sierpinski carpet* shown in Fig. 29(a). It is constructed as follows. Let us start with an occupied square: its mass M is 1, and so is its length L. Now the occupied square is replaced by 3×3 squares, of which the centre square is empty and the neighbouring eight squares are occupied, as shown in Fig. 29(a). The mass is now 8, not counting the empty square, and $L = 3$. This process is repeated again and again, with occupied squares replaced by 8 occupied and one empty centre square, and empty squares replaced by 9 empty squares.

Figure 29(a) shows the next step with $M = 64$ occupied squares and a length $L = 9$. At each iteration, L increases by a multiplicative factor 3 and M increases by a factor 8. After n iterations we have $L = 3^n$ and $M = 8^n$, and thus

$$M = L^D$$

with

$$D = \frac{\log 8}{\log 3}$$

This *fractal dimension* $D = 1 \cdot 893$ is nearly exactly equal to the $D = 1 \cdot 896$ for the incipient infinite cluster of percolation in two dimensions. A generalization of this to three dimensions (a cube is replaced by 20 full cubes and 7 empty ones) yields $D = \log 20 / \log 3 = 2 \cdot 73$, which is 8 per cent above the fractal dimension $2 \cdot 53$ of the incipient infinite cluster there.

If we consider the full squares as imitations of occupied sites on the infinite cluster, then (in contrast to the existence of 'red' bonds on that cluster)

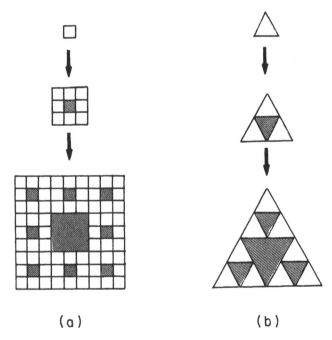

(a)   (b)

Fig. 29.   (a) Initial stages of the build-up of the Sierpinski carpet. Empty squares are shadowed. At each step of the iteration, the linear dimension L is enlarged by a factor 3 and the mass by a factor 8, since each occupied square is replaced by a 3×3 array of nine squares of which the centre square is empty. (b) Same for the Sierpinski gasket. Here the linear dimension is enlarged by $b = 2$, and the mass by a factor 3 at each iteration.

the Sierpinski carpet has an infinite *order of ramification*, i.e. one cannot cut an arbitrary piece of it by cutting a finite number of sites. It also does not have dangling ends. We shall cure these disadvantages below.

Another ordered recursive fractal, the *Sierpinski gasket*, is shown in Fig. 29(b). Here the basic building block is a triangle, and each iteration replaces it by a doubly sized triangle, with three full and one empty small triangles. After n iterations, $L = 2^n$, $M = 3^n$, and hence $M = L^D$ and

$$D = \frac{\log 3}{\log 2} = 1 \cdot 585$$

This seems to be a good approximation for the backbone of the infinite percolation cluster in two dimensions. A d-dimensional generalization uses hypertetrahedra, and yields

$$D = \frac{\log (d + 1)}{\log 2}$$

In three dimensions this yields $D = 2$, about 15 per cent higher than the backbone dimension for percolation.

Although the gasket has a *finite order of ramification*, equal to $(d + 1)$ (this is the smallest number of bonds you need to cut in order to separate an arbitrarily large piece of the structure), it has no singly connected bonds. It also has other details which differ from those of the percolation backbone. Nevertheless, many groups have used it since 1980 to study the behaviour of a large variety of physical problems on fractals.

Many of the above difficulties were resolved by a different model, proposed by Mandelbrot and Given (1984). They proposed the recursive construction shown in Fig. 30(a): one begins with a straight segment of unit length, and at each iteration one replaces it by eight segments. The length scale changes by a factor $b = 3$, and thus

$$D = \frac{\log 8}{\log 3} = 1 \cdot 893$$

This is the same for the Sierpinski carpet, and Fig. 30(b) indeed shows how the curve will fill the carpet asymptotically after an infinite number of iterations.

This curve has many more details which imitate the percolation incipient infinite cluster at p_c in two dimensions. Out of the 8 new bonds, 6 form the *backbone*. Out of these, 3 form the *minimal path*, 5 form the *maximal path* and 2 are *singly connected*. Further, the *resistance* between the ends is multiplied, after each iteration, by a factor

$$2 + \frac{1 \times 3}{1 + 3} = \frac{11}{4}$$

Thus,

$$D_B = \frac{\log 6}{\log 3} = 1 \cdot 631 \qquad D_{\min} = \frac{\log 3}{\log 3} = 1$$

$$D_{\max} = \frac{\log 5}{\log 3} = 1 \cdot 465 \qquad D_{SC} = \frac{\log 2}{\log 3} = 0 \cdot 631 \tag{93}$$

$$\tilde{\zeta}_R = \frac{\log (11/4)}{\log 3} = 0 \cdot 921$$

All these numbers are quite close to those of the percolation cluster in two dimensions.

The Mandelbrot–Given curve is easily generalizable: if we replace the two singly connected bonds by L_1 such bonds, the two dangling bonds by L_4 bonds, the $(1 + 3)$ bonds in the 'blob' by $(L_2 + L_3)$ (with $L_2 \leqslant L_3$), and use a generalized rescale factor b, then we have

$$D = \log(L_1 + L_2 + L_3 + L_4)/\log b$$

$$D_B = \log(L_1 + L_2 + L_3)/\log b$$

$$D_{\min} = \log(L_1 + L_2)/\log b$$

$$D_{\max} = \log(L_1 + L_3)/\log b \tag{94}$$

$$\tilde{\zeta}_R = \log\left(L_1 + \frac{L_2 L_3}{L_2 + L_3}\right)/\log b$$

$$D_{SC} = \log L_1/\log b$$

The choice $L_1 = 3$, $L_2 = 1$, $L_3 = 3$, $L_4 = 8$ and $b = 3$ yields

$$D = 2 \cdot 465 \qquad D_B = 1 \cdot 771 \qquad D_{\min} = 1 \cdot 262 \qquad D_{\max} = 1 \cdot 631$$

$$\tilde{\zeta}_R = 1 \cdot 203 \qquad D_{SC} = 1$$

all of which are surprisingly close to their counterparts for the percolation infinite cluster in three dimensions. As one increases d one has more and more singly connected bonds (L_1) and dangling bonds (L_4), and fewer and fewer bonds in the blobs (L_2, L_3). We leave it as a challenge to the reader to invent good generalized Mandelbrot–Given curves for $d = 4$ and 5.

Having invented a simple recursive fractal model, we can solve new problems on top of it, and hope that the results will not be far from their counterparts on the real random percolation cluster which they are supposed to imitate. For example, Blumenfeld *et al.* (1986) found that the multifractal exponents for the current distribution on the generalized Mandelbrot–Given curve are given by

$$y(q) = \frac{\log [L_1 + (L_2 L_3^{2q} + L_3 L_2^{2q})/(L_2 + L_3)^{2q}]}{\log b} \tag{95}$$

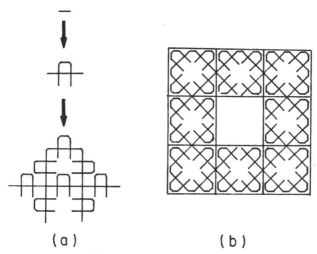

(a)   **(b)**

Fig. 30. (a) Initial stages of the Mandelbrot–Given curve. At each stage, the unit segment is replaced by 8 segments, with the new length scale $b = 3$. (b) Relation to the Sierpinski carpet.

The ambitious student is encouraged to check this result, which yields an excellent analytic approximant for these exponents. Equation (95) can now be used to estimate the 'noise' exponent, $y(2)$, for percolation in two and three dimensions, and the results agree very well with existing numerical estimates: $y(2)$ is slightly above $1/\nu$.

5.6. RENORMALIZATION GROUP FOR INTERNAL CLUSTER STRUCTURE

Another convenient analytical way to obtain results without heavy computation is to use approximate *small cell normalization group* schemes, like the ones described in Section 4.2. For resistor networks and similar problems, which involve *connectivity*, it is very convenient to use the scheme presented in Fig. 19 (see also Section 4.5). We now put a unit voltage between the lefthand side (sites A and D in Fig. 19(a) and the righthand side (sites C and F). The average resistance of the renormalized resistor is now easily calculated:

$$\frac{p'}{G'} = 1 \times p^5 + 1 \times p^4(1-p) + \tfrac{5}{3} \times 4p^4(1-p) + 3 \times 2p^3(1-p)^2$$
$$+ 2 \times 6p^3(1-p)^2 + 2 \times 2p^2(1-p)^3$$
$$= 4p^2 + 6p^3 - \tfrac{49}{3}p^4 + \tfrac{22}{3}p^5 \tag{96}$$

Thus, at $p^* = 1/2$, $1/G = 1 \cdot 917$, hence $\tilde{\zeta}_R = \log(1/G')/\log 2 = 0 \cdot 939$, close to the numerical result for two-dimensional percolation of $0 \cdot 975$.

Hong and Stanley (1983) also used the same scheme to obtain

$$D_{\min} = 1 \cdot 086 \qquad D_{\max} = 1 \cdot 285 \qquad\qquad (97)$$

We leave it as an exercise to use this scheme for deriving the multifractal exponents $y(q)$.

5.7. CONTINUUM PERCOLATION, SWISS-CHEESE MODELS, AND BROAD DISTRIBUTIONS

So far, our resistor network has contained only two types of resistors: those (unoccupied bonds) with infinite resistance (zero conductance), which have concentration $(1 - p)$, and those with unit conductance ($\sigma_b = 1/r_b = 1$). It is often the case that the finite conductances are not all equal to each other, but rather have a *distribution* $f(\sigma)$ ($f(\sigma)\,d\sigma$ is the fraction of 'good' bonds with conductance between σ and $(\sigma + d\sigma)$). As we saw in Section 5.4, a *narrow* distribution $f(\sigma)$ results with an unchanged resistance exponent $\tilde{\zeta}_R$, but with a multifractal behaviour of the resistance cumulants, or noise.

The situation is particularly interesting if the distribution $f(\sigma)$ is very *broad*. In such cases, the conductance of the whole network may be dominated by one 'bottleneck' bond which has a very low conductance.

As an example, consider the power-law distribution

$$f(\sigma) \propto \sigma^{-w} \qquad\qquad (98)$$

with $0 \leqslant \sigma \leqslant \sigma_{\max}$. As we discuss below, such distributions arise naturally in various *continuum models* of conduction or fluid–flow permeability in porous media (Feng *et al.*, 1987). In fact, if a bond is characterized by a 'width' δ, then its conductance can be shown to behave as

$$\sigma \propto \delta^{y+1} \qquad\qquad (99a)$$

and the exponent y depends on the nature of conductivity of interest (electrical or viscous fluid) and on the detailed shape of the 'bond' (see below). If the distribution of widths $P(\delta)$ approaches a constant as $\delta \to 0$, then the relation $P(\delta)\,d\delta = f(\sigma)\,d\sigma$ yields the relation

$$w = \frac{y}{y+1} \qquad\qquad (99b)$$

For $w < 0$, one can bound the total resistance $(1/G)$ both from above and from below by expressions in which each bond resistance $1/\sigma_b$ or conductance σ_b is replaced by its average $\langle 1/\sigma \rangle$ or $\langle \sigma \rangle$. At p_c, both these bounds behave as $L^{\tilde{\zeta}_R}$, and the conductivity scales as in the previous case. The situation becomes more interesting when $w > 0$. In this case, the links–nodes–blobs model turned out to be very useful in deriving a lower bound for the resistance. As we argued in Section 5.3, the resistance $1/G$ of a sample of size L at p_c is bounded from below by that of the singly connected bonds between the terminals,

$$\frac{1}{G} \geq \sum_{b}^{M_{SC}(L)} \left(\frac{1}{\sigma_b}\right) \tag{100}$$

For large L, $M_{SC} \propto L^{D_{SC}}$ is also large. For $w < 0$, the distribution $f(\sigma)$ has a finite average of $(1/\sigma)$, and the integral

$$\langle 1/\sigma \rangle = \int_0^{\sigma_{max}} d\sigma f(\sigma)/\sigma$$

converges. In this case, the righthand side of Eq. (100) may be replaced by $M_{SC}(L) \langle 1/\sigma \rangle$, and we recover the inequality of Eq. (80), $\tilde{\zeta}_R \geq D_{SC} = 1/\nu$. For $w > 0$, the above integral diverges. In this case, the sum on the righthand side of Eq. (100) will be dominated (and bounded from below by) the largest term in it, $1/\sigma_{min}$. Since $P(\delta)$ approaches a constant at small δ, one expects that the M_{SC} values of δ are distributed evenly between 0 and δ_{max}, and therefore that $\delta_{min} \propto \delta_{max}/M_{SC}$. Thus,

$$\frac{1}{G} \geq \frac{1}{\sigma_{min}} \propto \frac{1}{\delta_{min}^{y+1}} \propto M_{SC}^{y+1} \propto L^{D_{SC}(y+1)} = L^{D_{SC}/(1-w)} \tag{101a}$$

Writing $1/G \propto L^{\tilde{\zeta}_w}$, this implies that

$$\tilde{\zeta}_w \geq \frac{D_{SC}}{1-w} = D_{SC}(1+y) \tag{101b}$$

It was argued by Straley (1982), and confirmed by recent renormalization group arguments, that Eq. (101) becomes an equality when $D_{SC}/(1-w) \geq \tilde{\zeta}_R$, i.e. when $w > w_c = 1 - D_{SC}/\tilde{\zeta}_R$. In this range of w, there are many bonds with very small widths. However, when such a bond occurs inside a blob, there almost always exist other bonds in parallel to it, with much larger conductances. Therefore, the total resistance is dominated by the narrowest singly connected bonds.

As mentioned above, distributions like Eq. (98) occur in various continuum models of porous media. An illuminating example of such a model is the 'Swiss-cheese' model (Feng *et al.*, 1987). In three dimensions, this model is constructed by placing uniformly-sized spherical holes at random in a uniform transport medium. The holes are allowed to overlap with one another. The system supports transport only for hole volume fractions q below a threshold q_c, and (for $d \leq 6$) the conductivity for $q < q_c$ behaves as $\Sigma \propto (q_c - q)^\mu$, with

$$\mu = (d - 2 + \tilde{\zeta}_w)\nu$$

(See Eq. (78).)

The continuum Swiss-cheese model can be mapped onto a random network, in which the bonds represent 'necks' bounded by interpenetrating holes. A two-dimensional version of the model is shown in Fig. 31. As Feng *et al.* showed, the distribution of the widths of the necks $P(\delta)$ indeed

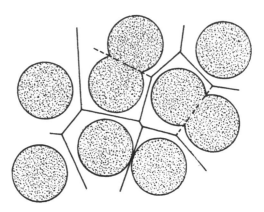

Fig. 31.   Swiss-cheese model in two dimensions. Straight lines show the bonds of the superimposed discrete network; dotted lines are the missing bonds. After Feng *et al.* (1987).

approaches a constant for $\delta \to 0$. They also showed that the electrical conductance of such necks obeys Eq. (99a), with $y = -1/2$ and $1/2$ for $d = 2$ and 3. In contrast, they found that the permeability of a viscous fluid through such necks has $y = 3/2$ and $5/2$ in $d = 2$ and 3.

Another interesting problem arises when no bonds are actually absent, but the range of bond resistances is very broad. Such distributions $f(\sigma)$ arise when the transport involves quantum mechanical tunnelling and/or thermal activation over barriers, with a wide range of barrier heights. A good way to estimate the conductance of the system, first proposed by Ambegaokar, Halperin and Langer (1971) is to identify the bond with the largest conductance, the bond with the next largest conductance, and so on. If we identify and mark these bonds consecutively, with decreasing conductance, we shall eventually reach a conductance value σ_c such that all the marked bonds (with conductances σ_b above σ_c) have exactly the percolation threshold concentration $p_c = \int_{\sigma_c}^{\sigma_{\max}} d\sigma \, f(\sigma)$. Beyond this point one has a spanning infinite cluster of such bonds. For sufficiently broad distributions, the conductance of this cluster turns out to yield an excellent estimate for that of the whole sample.

For concreteness, imagine that the bond conductance is given by

$$\sigma = \sigma_0 e^{-\Lambda\rho} \tag{102}$$

and ρ is uniformly distributed between 0 and 1. (For thermal hopping over barriers of energy ΔE, one has $\sigma \propto e^{-\Delta E/kT}$). In this case, the distribution function of σ is

$$f(\sigma) = \frac{1}{\Lambda\sigma} \qquad \sigma_0 e^{-\Lambda} \leqslant \sigma \leqslant \sigma_0$$

Note the similarity to Eq. (98), with $w = 1$. The broadness of the distribution is measured by Λ. Now

$$p_c = \int_{\sigma_c}^{\sigma_0} d\sigma \, f(\sigma) = \frac{1}{\Lambda} \ln \frac{\sigma_0}{\sigma_c} \quad \text{i.e.} \quad \sigma_c = \sigma_0 e^{-\Lambda p_c}$$

Such distributions are thus called 'logarithmically broad'. For $p > p_c$, we add bonds with $\sigma_1 < \sigma < \sigma_c$, and

$$p - p_c = \int_{\sigma_1}^{\sigma_c} d\sigma \, f(\sigma) = \frac{1}{\Lambda} \ln \frac{\sigma_c}{\sigma_1}$$

Using the links–nodes–blobs picture, the conductance of the total sample is given, according to Eq. (83), by

$$G(L) = (L/\xi)^{d-2} G(\xi) = L^{d-2} \Sigma$$

The resistance of a link of length ξ, $1/G(\xi)$, is bounded from below by that of the singly connected bonds, $1/G(\xi) \geqslant M_{SC}(\xi)\langle 1/\sigma \rangle$. Using

$$\langle 1/\sigma \rangle = \int_{\sigma_1}^{\sigma_0} d\sigma \, f(\sigma)/\sigma = \frac{1}{\Lambda} \left(\frac{1}{\sigma_1} - \frac{1}{\sigma_0} \right) \simeq \frac{1}{\Lambda \sigma_1}$$

and

$$M_{SC}(\xi) \propto \xi^{1/\nu} \propto (p - p_c)^{-1} = \frac{\Lambda}{\ln(\sigma_c/\sigma_1)}$$

we find

$$G(\xi) \leqslant G_1(\xi) \propto \sigma_1 \ln \left(\frac{\sigma_c}{\sigma_1} \right)$$

This upper bound will be optimized when $\ln(\sigma_c/\sigma_1) = 1$. Choosing this value of σ_1 (i.e. of p), and accepting the argument given above, that for sufficiently broad distributions G is given by this bound (blobs have much larger conductances), we find that

$$\Sigma \propto \sigma_c \Lambda^{(d-2)\nu} \propto \sigma_0 e^{-\Lambda p_c} \Lambda^{(d-2)\nu}$$

Although confirmed numerically for $d = 2$, at the moment it is not yet clear whether this heuristic result is correct for $d > 2$.

5.8. ELASTIC NETWORKS

In the good old times (about 1980) it was believed that the elastic behaviour of a disordered network like rubber is directly related to the electrical behaviour of the corresponding random resistor network. Just as we need a continuous chain of conductors to have an electrical current flowing, we need an elastically active network chain in the rubber to create some resistance against elongating it. The elastic modulus is proportional to the ratio of applied force and produced deformation, and therefore it was thought to be the analogue of the electrical conductivity and to vanish at p_c with the same exponent μ.

Unfortunately, reality is more difficult. For bond percolation on the square lattice, we replace each bond by an elastic string which tries to keep a unit distance between neighbouring lattice sites. Now even at $p = 1$ a square lattice is unstable against shear forces, and for $p < 1$ a triangular lattice can be unstable even if an infinite network of connected bonds exists. The threshold for finite elastic moduli is thus larger than that for connectivity. Such shifts of percolation thresholds are avoided if we supplement these central forces by bond-bending forces. Thus the deformation energy consists of two terms. A *spring energy* (as in Hooke's law), proportional to the squared length change, and a *bending energy*, proportional to the squared angular change. (Lengths and angles are measured between neighbouring sites and bonds, respectively.)

In this model the elastic behaviour becomes critical at the geometrical percolation threshold. However, elasticity and conductivity have different critical exponents. The change of an angle between two bonds is proportional not only to the applied force but also to the distance at which it is applied, i.e. it is given by the *torque*. This distance has no analogue in the conductance calculation and enters quadratically into the deformation energy. In the nodes-and-links picture, this distance is ξ. If one considers only the elastic response of the singly connected bonds, then the force constant (equal to the ratio of the force and the elongation of the 'spring', as in Hooke's law) of a typical link of linear size ξ is

$$k_\xi \propto \xi^{-2} M_{SC}(\xi)^{-1} \propto (p - p_c)^{2\nu + 1}$$

and therefore (using arguments like those which led to Eq. (78)) the elastic stiffness coefficient of the backbone behaves as $(p - p_c)^\tau$, with

$$\tau = (d - 2)\nu + 2\nu + 1 = d\nu + 1 \qquad (103a)$$

This value was identified by Kantor and Webman (1984) as a lower bound for τ. The above expression for τ is somewhat lower than the Monte Carlo result $\tau = 3 \cdot 96 \pm 0 \cdot 04$ of Zabolitzky *et al.* (1986) in two dimensions. Since elasticity introduces two extra factors of ξ, Sahimi (1986) and Roux (1986) conjectured that

$$\frac{\text{elasticity}}{\text{conductivity}} \propto \xi^{-2}$$

and therefore

$$\tau = \mu + 2\nu = (d + \tilde{\zeta}_R)\nu \qquad (103b)$$

in excellent agreement with the Monte Carlo result.

For $d > 6$, the above hyperscaling relations, which contain d, fail. In this case, the Kantor–Webman argument becomes exact, and $\tau = \mu + 2\nu = 4$. In practice, τ seems to be close to 4 in all dimensions $d \geqslant 2$.

For continuum percolation one can repeat all the arguments of the previous section, and recover

$$\tau = (d + \tilde{\zeta}_w)\nu \qquad (103c)$$

One should be warned that even the bond-bending model may not be realistic for rubber, where entropy effects may dominate.

FURTHER READING

First experiments on random resistor networks

Last, B.J. and Thouless, D.J., *Phys. Rev. Lett.*, **27**, 1719 (1971).

Monte Carlo measurements of conductivity

Derrida, B., Zabolitzky, J.G., Vannimenus, J. and Stauffer, D., *J. Stat. Phys.*, **36**, 31 (1984).
Gingold, D.B. and Lobb, C.J., *Phys. Rev. B*, **42**, 8220 (1990).
Kirkpatrick, S., *Rev. Mod. Phys.*, **45**, 574 (1973).
Normand, J.M., Herrmann, H.J. and Hajjar, M., *J. Stat. Phys.*, **52**, 441 (1988).

Series on conductivity

Adler, J., Meir, Y., Aharony, A., Harris, A.B. and Klein, L., *J. Stat.*, **58**, 511 (1990), and references therein.

Hall resistance and superconductors

Bergman, D.J., Duering, E. and Murat, M., *J. Stat. Phys.*, **58**, 1 (1990).
Normand, J.M. and Herrmann, H.J., *Int. J. Mod. Phys. C*, **1**, 207 (1990).

Internal structure of clusters and the links–blobs–nodes model.
See de Gennes, P.G., *La Recherche*, **7**, 919 (1976) and also:

Aharony, A., Gefen, Y. and Kapitulnik, A., *J. Phys. A*, **17**, L197 (1984).
Havlin, S., in: *Kinetics of Aggregation and Gelation*, edited by Family, F. and Landau, D.P. (Amsterdam: North Holland, 1984).
Herrmann, H.J. and Stanley, H.E., *J. Phys. A*, **21**, L879 (1988).
Pike, R., and Stanley, H.E., *J. Phys. A*, **14**, L169 (1981).
Skal, A.S. and Shklovskii, B.I., *Sov. Phys.–Semicond.*, **8**, 1029 (1975).
Stanley, H.E., *J. Phys. A*, **10**, L211 (1977).

Multifractals.
See Mandelbrot (1982) and Feder (1988) cited in Chapter 1, as well as:

Aharony, A., *Physica A*, **168**, 479 (1990).
Blumenfeld, R., Meir, Y., Aharony, A. and Harris, A.B., *Phys. Rev. B*, **35**, 3524 (1987).
de Arcangelis, L., Redner, S. and Coniglio, A., *Phys. Rev. B*, **31**, 4725 (1985).
Furuberg, L., Aharony, A., Feder, J. and Jøssang, T., in: *Time Dependent Effects in Disordered Materials*, edited by Pynn, R. and Riste, T. (New York: Plenum Press, 1987).
Rammal, R., Tannous, C., Breton, P. and Tremblay, A.M.S., *Phys. Rev. Lett.*, **54**, 1718 (1985).

Fractal models.
See Mandelbrot (1982) as well as:

Blumenfeld, R., Meir, Y., Harris, A.B. and Aharony, A., *J. Phys. A*, **19**, L791 (1986).
Gefen, Y., Aharony, A., Mandelbrot, B.B. and Kirkpatrick, S., *Phys. Rev. Lett.*, **47**, 1771 (1981).
Mandelbrot, B.B. and Given J.A., *Phys. Rev. Lett.*, **52**, 1853 (1984).

Renormalization group techniques.
See references after Chapter 4, and also:

Hong, D.C. and Stanley, H.E., *J. Phys. A*, **16**, L475, L525 (1983).

Continuum percolation

Ambegaokar, V., Halperin, B.I. and Langer, J.S., *Phys. Rev. B*, **4**, 2612 (1971).
Feng, S., Halperin, B.I. and Sen, P., *Phys. Rev. B*, **35**, 197 (1987).
Halperin, B.I., *Physica D*, **38**, 179 (1989).
Straley, J.P., *J. Phys. C*, **15**, 2343 (1982).

Elastic networks

Feng, S., and Sen, P.N., *Phys. Rev. Lett.*, **52**, 276 (1984).
Kantor, Y., and Webman, I., *Phys. Rev. Lett.*, **52**, 1891 (1984).
Roux, S., *J. Phys. A*, **19**, L351 (1986).
Sahimi, M., *J. Phys. C*, **19**, L79 (1986).
Zabolitzky, J.G., Bergman, D.J. and Stauffer, D., *J. Stat. Phys.*, **44**, 211 (1986).

Self-avoiding walks on percolation clusters

Meir, Y. and Harris, A.B., *Phys. Rev. Lett.*, **63**, 2819 (1989).

Viscous fingers on percolation clusters

Oxaal, U., Murat, M., Boger, F., Aharony, A., Feder, J. and Jøssang, T., *Nature*, **329**, 32 (1987).

Minimal path exponent of lattice animals

Havlin, S., Djordjevic, Z.V., Majid, I., Stanley, H.E. and Weiss, G.H., *Phys. Rev. Lett.*, **53**, 178 (1984).

CHAPTER 6
Walks, Dynamics and Quantum Effects

If you have studied equilibrium thermodynamics and/or statistical physics, and had some time left, you might have encountered the problem of *non-equilibrium effects*. For example, in equilibrium thermodynamics the temperatures of two objects in thermal contact, like Scotch and ice cubes, are the same. In non-equilibrium thermodynamics, we learn how quickly heat can diffuse from the warmer to the colder object, that is how long it takes to establish equilibrium. Similarly, in the scaling theory of phase transitions, shortly after static scaling laws were invented during the 1960s they were generalized to cover time-dependent or non-equilibrium effects like the thermal conductivity near the superfluid transition or the spin wave spectrum near the ferromagnetic Curie point (Hohenberg and Halperin, 1977). We now look for something similar in the percolation field.

To describe such non-equilibrium phenomena, often also called *transport properties*, it is often not sufficient just to know everything about the static behaviour. An additional time-dependent property is needed, too; then one can try to express other transport properties through this time-dependent property and the static quantities. For percolation, we take diffusion on random dilute networks as our basic transport property, and on this basis we will try to understand a variety of other properties. Specifically, we start with simulations and scaling theories for diffusion of a particle on a random network, called the 'ant in the labyrinth' and also discuss the distribution function of such walkers. We then proceed to discuss *fractons*, which describe localized dynamical excitations on dilute systems. The rest of the chapter discusses the accessible perimeter of clusters, relevant also for *self-affine diffusion fronts*, and *invasion percolation*, which is an example of a *self-organized critical* percolating system.

6.1. ANTS IN THE LABYRINTH

In Chapter 1 we mentioned the diffusion of a particle in a disordered network, a problem dubbed the 'ant in the labyrinth'. Now we want to look at this problem from a more theoretical point of view to see whether we can apply what we have learned in the meantime. First let us repeat the definition of this kinetic process. At every time step, the diffusing particle, called the ant, selects randomly one of its nearest neighbour sites. If that site is occupied ('permitted', probability p), it moves there; if the neighbour is empty ('prohibited', probability $1 - p$), it stays where it is. That process is repeated again and again, and averaged over many different ants running through many different lattices. To avoid any X-rating of this book, ant–ant interactions are ignored. We are interested in the average distance R which the ants travel as function of the time t; t is the number of time steps (jump attempts) made by the ant.

We have already said in Chapter 1 that for p far above p_c one observes normal diffusion, $R^2 \propto t$, for large times, whereas for p far below p_c, R approaches a constant for long times. From what we have learned in the preceding chapters we can assume that for all concentrations above the threshold one type of behaviour (diffusion) will dominate asymptotically, whereas for all $p < p_c$ the other behaviour (finite asymptotic distance) will be valid. A third type, i.e. anomalous diffusion, with $R \propto t^k$, will govern the asymptotic distance right at the critical point. We will now study in greater detail the behaviour very close to p_c where these three different laws have to merge.

If $P_i(t)$ is the conditional probability that the ant is at site i and time t, and if we start with the ant at the origin site $i = 0$ at time $t = 0$ ($P_0(0) = 1$ and $P_i(0) = 0$ for $i \neq 0$), then we are interested to find how the probability 'spreads' with time. In the above example, if all the z sites neighbouring the origin are occupied, then after one step the probability to be in any of them will be $1/z$, while all other $P_i(1)$ will be zero. If two of the neighbouring sites are empty, the ant will stay at the origin with probability $P_0(1) = 2/z$. More generally, $P_i(t)$ obeys a 'master equation',

$$P_i(t + 1) - P_i(t) = \sum_j \left[\sigma_{ji} P_j(t) - \sigma_{ij} P_i(t) \right] \qquad (104a)$$

where σ_{ji} is the probability for the ant to hop from site j to its nearest neighbour site i in one time step. The first term in Eq. (104a) arises from hops into site i, and the second term corresponds to hops out of this site.

For the example described above, $\sigma_{ji} = 1/z$ if site i is occupied, and $\sigma_{ji} = 0$ if site i is empty. Such an ant is called 'blind', or 'drunken', since it is not clever enough to identify the occupied neighbours. A more intelligent species, called 'myopic' ants, choose only among the occupied neighbours. These ants have $\sigma_{ji} = 1/z_j$, where z_j is the number of *occupied* neighbours of site j. Numerical simulations and rigorous arguments now agree that both species converge to practically the same asymptotic behaviour (Harris *et al.*, 1987).

At long times, we can replace the lefthand side of Eq. (104a) by

$$\frac{dP_i}{dt} = \sum_j \left[\sigma_{ji} P_j(t) - \sigma_{ij} P_i(t) \right] \qquad (104b)$$

For a *finite* cluster, one eventually reaches a *stationary* state, in which $P_i(t)$ no longer varies with time. The ant keeps hopping back and forth, but its average probability to be found at any site i remains time-independent, P_i(stationary). Setting $dP_i/dt = 0$, we see that P_i(stationary) must solve a set of linear equations. For the 'blind' ant, with $\sigma_{ji} = \sigma_{ij} = 1/z$, the solution requires that all the P_i are equal to each other, i.e.

$$P_i(\text{stationary}) = \frac{1}{s}$$

This implies that all the sites are equally probable, or that every site is visited equally frequently by the ant. For the 'myopic' ant, P_i(stationary) $\propto z_i/s$.

The fact that all the sites of a cluster have 'equal rights' means that asymptotically the distance R for one finite cluster is the average distance between two cluster sites. This average in turn was our cluster radius R_s defined in Eq. (45). Now we have to average over different cluster masses s. Initially an ant parachutes to each occupied site with the same probability. The probability of a site belonging to a cluster containing s sites is $n_s s$, and then the asymptotic distance is R_s. Usually one averages over the squared distances and then gets from the sum over all cluster sizes:

$$R^2(t = \infty, p < p_c) = \sum_s n_s s R_s^2$$
$$\propto (p_c - p)^{\beta - 2\nu} \tag{105}$$

The critical exponent in this result is derived in the same way that we derived the exponent for Eq. (31). We see that not all lengths are proportional to the correlation length ξ: the exponent is not simply 2ν. (Had we averaged over R_s instead of R_s^2, we would have got a critical exponent $\beta - \nu$; see Mitescu and Roussenq (in *Percolation Structures and Processes*, cited in Chapter 1) for data.) In three dimensions, the exponent $\beta - 2\nu$ is about $-1\cdot34$. Monte Carlo determinations first gave much higher values (Mitescu and Roussenq), then agreed better with theory (Pandey *et al.*, 1984). Let us thus hope that we understand the behaviour below the threshold; what about above and at p_c?

For $p = 1$ we showed in Chapter 1 that $R^2 = t$ exactly for all t, not only asymptotically. Generally one calls any law $R^2 \propto t$ a diffusion law, and we denote the proportionality constant as the diffusivity $\mathcal{D}$:

$$R^2 = \mathcal{D}t$$

for long times t. Thus for $p = 1$ we have diffusion with $\mathcal{D} = 1$, even for short times. (The usual definition of $\mathcal{D}$ differs from ours by a trivial d-dependent factor, which we ignore to have $\mathcal{D}(p = 1) = 1$.) For p slightly below unity but still far above p_c we have some holes in the lattice which slow down the ant but still let it diffuse everywhere. Thus, the diffusion law should still be valid, only with a reduced diffusivity $\mathcal{D} = \mathcal{D}(p)$. For p below p_c diffusion is impossible since now $R^2(t)$ is bounded by Eq. (105). Thus for very large times R cannot increase as $t^{1/2}$, and the diffusivity is zero. Thus $\mathcal{D}(p)$ goes to zero if p approaches p_c from above.

Fortunately, no new critical exponent is needed to describe how $\mathcal{D}$ vanishes near the percolation threshold. De Gennes (1976) (see Chapter 1) pointed out that the diffusivity $\mathcal{D}$ is proportional to the conductivity Σ of random resistor networks:

$$\mathcal{D} \propto \Sigma \quad \text{or} \quad R^2 \propto \Sigma t \quad (t \to \infty) \tag{106a}$$

This equation is simply a manifestation of Einstein's result from the beginning of this century, that in statistical physics the diffusivity is proportional to the mobility. The mobility, on the other hand, is the ratio of the velocity to the applied force. For the electrons in the copper parts of a random resistor network, the applied force is proportional to the electric field, that is to the voltage. The average velocity of the electrons is proportional to the electric current they produce. Therefore their mobility is proportional to the ratio of current to voltage, that is to the conductivity of the network. Thus Eq. (106a) is basically due to Einstein and therefore needs no further proof. Close to the percolation threshold we recover our conductivity exponent:

$$\mathcal{D} \propto (p - p_c)^{\mu} \tag{106b}$$

How can we combine the two so seemingly different results (Eqs. (105) and (106)) into one consistent theory? Having studied Eqs. (33), (52) and (55) the reader will have no difficulty in recognizing that scaling can again be applied to the distance R depending on two variables $1/t$ and $p - p_c$ both going to zero. The general statements discussed after Eq. (52) also apply here. (A warning: The analogy with our earlier results is closest if we regard the time t as function of the distance R and replace the system length L by the ant distance R in Eq. (55). But that last step would be incorrect, since R according to Eq. (105) does not scale as ξ.) With two suitable exponents x and k and a scaling function $r(z)$ we assume

$$R = t^k r[(p - p_c)t^x] \tag{107}$$

For p above p_c and sufficiently long times and distances we must recover the diffusion law of Eq. (106); thus for large positive arguments z the scaling function $r(z)$ is proportional to $z^{\mu/2}$ in order to be consistent with Eq. (106). Then

$$R \propto t^k (p - p_c)^{\mu/2} t^{\mu x/2} \propto t^{k + \mu x/2} \mathcal{D}^{1/2}$$

We will also need $R \propto t^{1/2}$ in this regime, which requires $k = (1 - \mu x)/2$. (Our R is the root mean square distance, that is the square root of the averaged R^2.)

On the other hand, for p below p_c we must recover the result given by Eq. (105) that R varies as $(p - p_c)^{-\nu + \beta/2}$ independently of t for sufficiently long times. The scaling function $r(z)$ for $z \to -\infty$ thus must vary as $(-z)^{-k/x}$ in order that t cancels out: $R \propto (p_c - p)^{-k/x}$. Equation (105) now requires this exponent k/x to equal $\nu - \beta/2$, or $k = (\nu - \beta/2)x$. Equating these two expressions for k we get $1 - \mu x = (2\nu - \beta)x$, or

$$x = \frac{1}{2\nu + \mu - \beta} \tag{108a}$$

from which

$$k = \frac{\nu - \beta/2}{2\nu + \mu - \beta} \tag{108b}$$

follows.

We have thus derived the two exponents x and k entering the scaling assumption (107). If we simulate the ant right at the critical point, we get an 'anomalous diffusion' exponent k smaller than the usual $1/2$ from Eq. (107):

$$R \propto t^k \propto t^{(\nu - \beta/2)/(2\nu + \mu - \beta)} \tag{109}$$

for long times t and $p = p_c$. This anomalous exponent k is about $0 \cdot 33$ in two dimensions and close to $0 \cdot 2$ in three dimensions. Numerous numerical tests have confirmed these predictions with reasonable accuracy. Figure 32 shows three-dimensional data whereas Fig. 5 (Chapter 1) has already given two-dimensional results.

We see from the argument z of the scaling function $r(z)$ in Eq. (107) that there is a characteristic time in our relation between R and t. For times much smaller than $|p - p_c|^{-1/x}$ but much larger than unity one has anomalous diffusion, Eq. (109), whereas for much longer times one either observes normal diffusion, Eq. (106), or a constant distance, Eq. (105). The characteristic time $\propto |p - p_c|^{-1/x} = |p - p_c|^{\beta - 2\nu - \mu} \propto |p - p_c|^\beta \xi^2 / \mathcal{D}$ separates the regime of anomalous diffusion from the more usual behaviour.

You may wonder why this characteristic time is not simply proportional to $\xi^2 / \mathcal{D}$, the time needed for an ant to diffuse over a distance of the order of

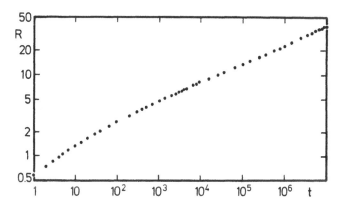

Fig. 32. Root mean square distance R travelled by a diffusing particle ('ant') in t step attempts on a simple cubic lattice at its percolation threshold.

the correlation length ξ. Where does the additional factor $|p - p_c|^\beta$ come from? The reason is the same as that for Eq. (105). It comes from averaging over all cluster sizes. Percolation properties are usually derived as being the sum of the contributions from single clusters. Thus a more microscopic scaling theory (Gefen *et al.*, 1983; Ben Avraham and Havlin, 1982) starts with a generalization of Eq. (107) to the distance R travelled by an ant during the time t in a cluster containing s sites: $R = R(s, t, p)$ instead of the above $R(t, p)$. Summation over all s, as in Eq. (105), then has to give Eq. (107) and the other results above.

If we now look for p slightly above p_c at very large clusters only, with s much larger than $|p - p_c|^{-1/\sigma}$, and with cluster radii much larger than both the correlation length and the distance travelled by the ant, then the ant will not feel the cluster boundaries. Thus it will behave as if it were on the infinite cluster. In the homogeneous regime, when $R \gg \xi$, the ant diffuses as usual: $R^2 = \mathscr{D}'t$. This new diffusivity $\mathscr{D}'$, however, is not the same as our diffusivity $\mathscr{D}$ defined above for the case where the ant starts running on an arbitrary occupied site, not necessarily on the infinite cluster. If it starts anywhere on an occupied site, then with a probability P/p it starts on the infinite cluster, and with the probability $1 - P/p$ it starts on a finite cluster. For the strength P of the infinite cluster is the probability that an arbitrary lattice site, occupied or empty, belongs to the infinite cluster. Only an ant on the infinite cluster can contribute to a distance increasing with time ($R^2 = \mathscr{D}'t$); the other ants add only a finite amount to the distance. Thus, if we average over all occupied sites as starting points of the random walk we get for sufficiently long times

$$\mathscr{D}t = R^2 = (P/p)\mathscr{D}'t$$

The diffusivity $\mathscr{D}$ in the whole lattice and the diffusivity $\mathscr{D}'$ in the infinite network or very large cluster are thus related by

$$\frac{\mathscr{D}}{\mathscr{D}'} = \frac{P}{p} \qquad (110)$$

an equality not restricted to the region very close to the percolation threshold.

Now we see that our characteristic time $|p - p_c|^{-1/x}$ discussed above is nothing but the time $\xi^2/\mathscr{D}'$ the ant needs to travel the distance ξ if it diffuses, with diffusivity $\mathscr{D}'$, on an infinite or very large cluster. For close to the threshold we have $\mathscr{D}' \propto \mathscr{D}/(p - p_c)^\beta$, and thus we have explained the unexpected factor $|p - p_c|^\beta$ in our characteristic time. This factor is not a violation of scaling but merely indicates that different types of averages may have different critical exponents.

For a theory of diffusion on finite clusters with radius R_s it is not sufficient to assume

$$R = R_s r[(p - p_c)t^x]$$

in analogy with Eq. (107). Instead the scaling function r depends also on the scaling variable $(p - p_c)s^\sigma$. We do not go into these details (Gefen *et al.*, 1983), since we want to restrict this book to scaling functions of a single variable only.

Monte Carlo studies have been made both with ants running everywhere where permitted (e.g. Pandey *et al.*, 1984), or being restricted to the infinite cluster (e.g. Ben Avraham and Havlin, 1982). In both cases, they confirm the above values for k and k'.

Another way to understand these scaling relations is to consider diffusion on the infinite cluster only (Gefen *et al.*, 1983). If $R \gg \xi$, then the ant 'sees' a *homogeneous* cluster, and we expect that $R^2 = \mathscr{D}'t$, with $\mathscr{D}' \propto (p - p_c)^{\mu - \beta} \propto \xi^{-\theta}$ and $\theta = (\mu - \beta)/\nu$. If $R \ll \xi$, the ant moves on a fractal structure, and we expect anomalous diffusion, with $R \propto t^{k'}$. Since both behaviours should match each other at $R = \xi$, we find that

$$k' = \frac{\nu}{2\nu + \mu - \beta} = \frac{1}{2 + \theta} \tag{111}$$

This anomalous diffusion exponent is about $0 \cdot 35$ in two and $0 \cdot 27$ in three dimensions. As explained in Section 1.4, $1/k'$ may be identified with the fractal dimensions of the random walk on the incipient infinite cluster at p_c. It is often denoted by $d_w = 1/k' = 2 + \theta$.

Note that the fractal dimension d_w is different for different classes of walkers: it is equal to $1/k'$ or to $1/k$ for walks restricted to the infinite cluster or averaged over all clusters. Using Eqs. (54b) and (78), Eq. (111) may be written as

$$d_w = \frac{1}{k'} = 2 + \frac{\mu - \beta}{\nu} = D + \tilde{\zeta}_R \tag{112a}$$

The resulting value of d_w thus depends on the fractal dimension D of the structure on which the ant moves: if the ant is restricted to move only on the backbone, and to avoid dangling bonds, then the fractal dimension of its anomalous walk becomes

$$d_{wB} = D_B + \tilde{\zeta}_R \tag{112b}$$

Yet another way to look at the random walk between two points on the infinite cluster is to note that, since the ant steps only on occupied sites, the walk really depends only on the *topology* of the cluster and not on the particular way in which the cluster is embedded on the lattice. Therefore, it is more natural to measure the distance travelled by the ant in terms of the minimal path, or chemical distance (see review by Havlin and Ben Avraham, 1987). Replacing R by $l_{min} \sim R^{D_{min}}$, the relation $t \sim R^{d_w}$ is replaced by $t \propto l_{min}(R)^{d_{wm}}$, and

$$d_{wm} = \frac{d_w}{D_{min}} \tag{113}$$

(In some respect, d_{wm} is more 'basic' than d_w. For example, random walks on top of polymer chains always have $d_{wm} = 2$, as in one dimension, even though d_w and $D = D_{min}$ vary depending on the dimension of space in which these polymers are embedded).

Scaling assumptions like Eq. (107) have counterparts in thermal critical phenomena (Hohenberg and Halperin, 1977). There, one encounters characteristic times which diverge near the thermal critical point as ξ^z, where ξ is the thermal correlation length. Clearly, z is the analogue of our d_w.

Finally, we note that if we replace the 'good' bonds by superconducting ones (infinite conductance, or zero hopping time) and the 'bad' bonds by 'normal' ones ($\sigma_{ij} = 1/z$ in Eq. (104)), then we can describe this normal-to-superconducting transition by a new kind of random walkers, called 'termites'. More details on this can be found in the reference list.

Diffusion of ants is also a natural way to introduce *directed percolation* (see Duarte (1990) for a recent study). If we orient the square lattice in the directions of north, east, south and west, then we may assume that an ant may walk only in the directions north and east, and never in the southern or western directions. Now it is more difficult to find an infinite network of allowed directions, if the lattice is still occupied randomly. Thus the percolation threshold shifts upwards, and also the critical exponents are changed. We now have two correlation lengths: $\xi' \propto |p - p_c|^{-\nu'}$ for correlation along the preferred direction (north–east), and $\xi \propto |p - p_c|^{-\nu}$ perpendicular to this direction. The scaling law for the strength of the infinite cluster (exponent β) and the mean cluster size (exponent γ) (Eq. (53)) now reads $\gamma + 2\beta = (d - 1)\nu + \nu'$ in d dimensions. For $d = 2$, the numerically determined exponents are very close to $\beta = 199/720$, $\gamma = 41/18$, $\nu = 79/72$, $\nu' = 26/15$. More difficult is the general case when the ant travels north merely with a higher probability than in the other directions, but can still walk backwards.

6.2. PROBABILITY DISTRIBUTIONS

The distance $R(t)$ has been defined as the root mean square distance travelled by the ant at time t. Instead of restricting ourselves to this average, we may be interested in the complete distribution function of these distances, $P(r, t)$. In the discrete lattice case, $P(r, t)$ is the probability for the ant to reach a distance r after t steps. For large t or small lattice constant we may use a continuum version, when $P(r, t)d^d r$ is the probability of finding the ant in a volume $d^d r$ around r. On homogeneous ordered systems, one can show (e.g. by solving Eq. (104)) that $P(r, t)$ is exactly a *Gaussian*, i.e.

$$P(r, t) = \frac{1}{(2\pi \mathscr{D}t)^{d/2}} \, e^{-r^2/(2\mathscr{D}t)} \tag{114}$$

The prefactor, $(2\pi \mathscr{D}t)^{-d/2}$ comes simply from the normalization, which requires that the integral of $P(r, t)$ over all r (in d dimensions) equals unity.

It is interesting to note that the probability to return to the origin is given by

$$P(0, t) = (2\pi \mathscr{D}t)^{-d/2} \propto R(t)^{-d}$$

This probability is thus inversely proportional to the volume of the region visited by the ant. A heuristic argument for this result is that within this volume the ant visits every site many times. Therefore, the probability to visit each of them, and particularly the origin, is equal to one over their number.

Using Eq. (114), we can now calculate all the moments of the distance r at time t, and show that they obey simple (rather than multifractal) scaling,

$$\langle r^{2q} \rangle \propto R(t)^{2q} \propto t^q$$

Equation (114) is very similar to Eq. (23), for $n_s(p)$: it has a power-law prefactor, and an exponential cutoff. Following the arguments that led us from Eq. (23) to Eq. (33), it is tempting to generalize Eq. (114) for diffusion on the infinite cluster at p_c, and to write

$$P(r, t) = t^{-d_s/2} f(r^{d_w}/t) \tag{115}$$

The argument r^{d_w}/t is chosen so that the root mean square average $R(t)$ will scale as $t^{k'} = t^{1/d_w}$, and the moments $\langle r^{2q} \rangle$ will scale as R^{2q}. The prefactor, which is related to the probability to return to the origin, $P(0, t)$, can again be determined from normalization. Since the walk is restricted to move on the infinite cluster, whose mass within a region of linear size $R(t)$ scales as $R(t)^D$, we expect that $P(0, t) \propto R(t)^{-D} \propto t^{-D/d_w}$, and hence

$$d_s = \frac{2D}{d_w} \tag{116}$$

This new exponent d_s is called the *fracton* or the *spectral* dimension. The term was introduced by Alexander and Orbach (1982) and by Rammal and Toulouse (1983) in connection with the number of frequency dependent excitations on fractals. We shall say more on this in the next section. Using the limited information available in 1982 on the values of D and d_w (via their relations to the exponents ν, β and μ), Alexander and Orbach noted that d_s is very close to 4/3 in *all dimensions* $d \geqslant 2$. In fact d_s equals 4/3 exactly for $d > 6$. As a result, many authors attempted to check the daring 'Alexander–Orbach conjecture', that d_s is *always* equal to 4/3. If this were true, it would imply an exact expression of the dynamic exponent μ in terms of the static ones, β and ν. Unfortunately, the existing accurate values at $d = 2$ yield $d_s = 1 \cdot 32$, and seem to exclude the value 4/3. Similar small deviations arise in $d = 6 - \varepsilon$ dimensions. However, the Alexander–Orbach conjecture is certainly an excellent approximant for most practical purposes concerning walks on the infinite cluster.

As in our discussion for $n_s(p)$, and in view of the exponential decay in Eq. (114), we next assume that the scaling function $f(x)$ in Eq. (115) decays exponentially for large x, i.e.

$$P(r, t) \propto t^{-d_s/2} \exp[-A(r^{d_w}/t)^\alpha] \tag{117}$$

There exist several ways to identify the exponent α (Aharony and Harris, 1990). Instead of giving you these exact but complicated arguments, we present here a simple heuristic one: consider the maximal value of r after t discrete steps. This value is reached when t equals the minimal path on the cluster, i.e. $t \propto r^{D_{min}}$. Since each step of the ant has some finite probability K (equal to $1/z$ for the 'blind' ant), the probability for reaching this point is of order $K^t = e^{-t|\ln K|}$. On the other hand, Eq. (117) yields a probability of order $\exp[-A(r^{d_w}/r^{D_{min}})^\alpha]$. Thus $D_{min} = \alpha(d_w - D_{min})$, or (using Eq. (113)),

$$\alpha = \frac{D_{min}}{d_w - D_{min}} = \frac{1}{d_{wm} - 1} \tag{118a}$$

This value of α applies when we start with a given *specific realization* of the cluster, and we consider all the random walks which start at a fixed origin on that cluster. Such walks, and their corresponding distribution $P(r,t)$, are called 'typical'. If we move the origin over all the sites of the cluster, and repeat the walks on all possible clusters, then we end up with a *distribution of the distributions* $P(r,t)$. Bunde *et al.* (1990) recently showed that this distribution is multifractal. A signal of this multifractality is reflected by the fact that since the average over all distributions $\langle P(r,t)\rangle$ also contains a configuration in which the minimal path is a straight line, with $r = t$, it is expected that although $\langle P(r,t)\rangle$ is still given by Eq. (117), the exponent α is now replaced by

$$\alpha_{av} = \frac{1}{d_w - 1} \tag{118b}$$

Equation (117) can be used, in analogy to Eq. (B1) in Appendix B, to derive a *Flory approximant* for self-avoiding walks on the infinite cluster at p_c. In this case, the self-avoiding walk is limited to move only on the backbone. Therefore, one must minimize the energy

$$E_s = A\left(\frac{R_s^{d_{wB}}}{s}\right)^{\alpha_B} + B\frac{s^2}{R_s^{D_B}}$$

where d_{wB} is given by Eq. (112b) and $\alpha_B = D_{min}/(d_{wB} - D_{min})$. The result is $s \propto R^{D_{SAW}}$, with

$$D_{SAW} = \frac{D_B + \alpha_B d_{wB}}{2 + \alpha_B}$$

The resulting approximate values turn out to be very close to available numerical measurements.

6.3. FRACTONS AND SUPERLOCALIZATION

The righthand side of the master equation Eq. (104), is very similar to Kirchhoff's equations, Eq. (74). This should not be surprising, in view of the equivalence between Σ and D, Eq. (106). The analogy may be made more complete if we connect each site on the resistor network via a capacitor C to the ground, and allow for a time-dependent external current, I_i. Equation (74) now becomes

$$C \frac{dV_i}{dt} = \sum_j \sigma_{ij}(V_j - V_i) - I_i \tag{119}$$

Noting the symmetry $\sigma_{ij} = \sigma_{ji}$, this equation is clearly very close to Eq. (104).

Equation (119) should be used when the external current is alternating (a.c.), with frequency ω. In that case we expect the conductivity of the cluster at p_c to depend on ω. Since ω is an inverse time, it is related to a length

$$L(\omega) \propto \omega^{-k'} \tag{120}$$

such that the electrons in the network move back and forth within a range $L(\omega)$. If $L(\omega) \ll L$ and $L(\omega) \ll \xi$, then the response of the network should not depend on its size L or on ξ, and we expect $L(\omega)$ to replace L in determining the conductivity in Eq. (73b), i.e. $\Sigma \propto L(\omega)^{-\mu/\nu} \propto \omega^{k'\mu/\nu}$. If $L(\omega) \gg L$, then we return to $\Sigma \propto L^{-\mu/\nu}$. In general, this implies that at p_c we have the crossover form

$$\Sigma = L^{-\mu/\nu} g(L/L(\omega)) \tag{121}$$

Equations (104) and (119) are similar to those which arise in a network of elastic springs, as in Hooke's law. Newton's second law then assumes the form

$$m \frac{d^2 u_i}{dt^2} = \sum_j K_{ji}(u_j - u_i) + F_i \tag{122}$$

where u_i and F_i are the displacement of and the external force on site i, whose mass is m. K_{ij} is the force constant of the spring connecting sites i and j. It was this similarity, which ignored bond-bending forces, that led to the mistaken belief that elasticity and electricity are equivalent (Section 5.8). However, Eq. (122) may still describe some of the vibrational excitations of the network. On regular networks, these excitations are called *phonons*. They have the form of waves, with frequency ω and wavelength λ which are related to each other via the dispersion relation

$$\omega = \frac{2\pi c}{\lambda} \tag{123}$$

where c is the sound velocity.

On dilute fractal networks, the solutions to Eq. (122) are more complicated. Instead of waves, which are spread over the whole sample, a perturbation at the origin turns out to decay with distance, i.e. to remain localized. Such excitations are called *fractons*. If the perturbation is periodic, with frequency ω, then we expect the decay length $L(\omega)$ to depend on ω, via a power law. Since Eq. (122) has two time derivatives, and Eq. (104) has only one time derivative, it follows that now we must replace ω by ω^2, and

$$L(\omega) \propto \omega^{-2k'} \tag{124}$$

The number of sites affected by such a perturbation is of order $L(\omega)^D \propto \omega^{-2k'D} = \omega^{-d_s}$. A comparison with the phonon analogue $\lambda^d \propto \omega^{-d}$ led to the name 'fracton dimensionality' for d_s. The number of such excitations with ω between ω and $\omega + d\omega$, is called the *density of states* $N(\omega)\,d\omega$, and one finds that $N(\omega) \propto \omega^{d_s-1}$.

Similarly to our discussion of the scaling forms of $n_s(p)$ and of $P(r, t)$, we can now expect the fracton perturbation to decay as

$$u(r, \omega) \propto \exp[-A(r/L(\omega))^a] \tag{125}$$

Since the exponent a often turns out to obey $a \geqslant 1$, this is sometimes called *superlocalization*. Arguments similar to those which led to Eq. (118) show that for a typical fracton,

$$a = D_{\min} \tag{126}$$

The average over all configurations $\langle u(r, t) \rangle$ decays exponentially, with $a = 1$.

The reader should have enough experience now to realize that when $p > p_c$ we should anticipate a crossover from fractons (at short distances or high frequencies) to phonons (large distances, low frequencies). Equation (124) may thus be extended (for $L \gg L(\omega), \xi$) into

$$\omega = L(\omega)^{-1/2k'} f(L(\omega)/\xi)$$

The requirement that $\omega \propto 1/L(\omega)$ for $L(\omega) \gg \xi$, coming from Eq. (123), now yields $f(x) \propto x^{-1+1/2k'}$ for $x \ll 1$, hence

$$c \propto \xi^{1-1/2k'} \propto (p - p_c)^{(\mu-\beta)/2} \propto (\mathscr{D}')^{1/2} \tag{127}$$

and the sound velocity decreases as p_c is approached.

There are many other cases in which one ends up with *linear equations* like Eqs. (104), (119) and (122). In Heisenberg ferromagnets, each occupied site has a magnetic moment which is free to rotate in space, and one finds spin-wave excitations. The equation of motion of these waves is equivalent to Eqs. (104) (for the 'myopic' case) and (119), with a single time derivative. In this case, Eq. (123) is replaced by the quadratic form

$$\omega = K(2\pi/\lambda)^2$$

and the (super-)localized excitations decay over a length $L(\omega) \propto \omega^{-k'}$, Eq. (120). For $L(\omega) \gg \xi$, the stiffness constant K scales exactly as the diffusion coefficient $\mathscr{D}' \propto (p - p_c)^{\mu-\beta}$.

Another case of great interest is that of quantum percolation. The Schrödinger equation for a quantum particle on a lattice, with hopping coefficients σ_{ji}, has the form

$$E\psi_i = \sum_j \sigma_{ji}\psi_j \tag{128}$$

On regular lattices this yields Bloch waves. However, on the percolation cluster at p_c it yields superlocalized wave functions. Localization of electrons in random media has been a very active field of research in recent years. Even a weak randomness, with all σ_{ij} being non-zero, may lead to a localization of the wave function, resulting from a destructive interference of the random phases of the complex wave functions ψ_i. As a result, the quantum wave functions remain localized even for concentrations above p_c, up to a new quantum threshold p_q (see, e.g., Meir *et al.* (1989) and references therein).

6.4. HULLS AND EXTERNAL ACCESSIBLE PERIMETERS

So far we have discussed only diffusion of particles *on* the percolating clusters. In many applications, one is interested in particles which diffuse *on the medium* which surrounds the cluster, and occasionally hit the cluster from the *outside*. To study this case, we need to solve the master equation Eq. (104), with new boundary conditions: $P_i(t) = 1$ on a far-away source of particles (e.g., a very large sphere), and $P_i(t) = 0$ on the cluster (assuming that particles are not allowed to step on it). One may then ask about the probability that a specific random walker, released from the far-away source, will hit the cluster at a particular point on it. In two dimensions, there are many internal screened sites, which will never be hit. The remaining sites, which have non-zero probabilities of being hit, form the *accessible external perimeter*. The hitting probabilities of sites on this perimeter turn out to have a very broad range: some sites are hidden inside deep screened fjords, and some sit on sharp external tips. In fact, the distribution of these hitting probabilities turns out to be multifractal. This and similar distributions are currently investigated by many groups who are interested in aggregation.

In this section we limit ourselves to the number of sites on the accessible external perimeter in two dimensions. As we shall see, there exist several ways to identify such a perimeter. We start with the *hull*, which contains sites *on* the cluster that neighbour vacant sites which are connected to the outside. The meaning of this connection will become clear below.

We now describe a simple numerical algorithm, which probes the hull as well as other interesting ingredients of the cluster (Grossman and Aharony, 1986, 1987). Given a finite cluster on the square lattice, at p_c, identify two endpoints on it, e.g. the points with the largest (smallest) y coordinate and—among these—with the largest (smallest) x coordinate. The Euclidean distance between these endpoints may serve as a measure for the linear size of the cluster, and it indeed turns out to scale as $s^{1/D}$ (Eq. (48)). The external perimeter is now probed using a *biased walk*. (Such walks are very effective if you want to find your way out of a labyrinth.) Beginning at the lower endpoint, the walker attempts to move to its nearest neighbour on the left. If that site is vacant, the walker tries to move upwards. If that neighbour is vacant, the walker moves to the right. The procedure is now repeated iteratively, and

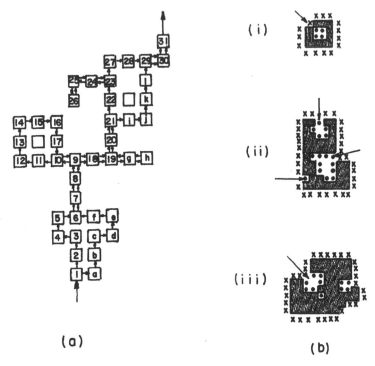

(a)

(b)

Fig. 33. (a) Example of left-turning (numbers) and right-turning (letters) biased walks, covering the hull sites. (b) The accessible perimeter E_1, E_2 and E_3 on the square lattice: shaded squares represent occupied sites; crosses = accessible perimeter sites; solid circles = screened inaccessible perimeter sites. (i) The arrow points at the unpenetrable gate from the external perimeter E_1 (cross) into a screened part. (ii) The arrow points at the unpenetrable entrances for particles larger than unity. The E_2 sites are denoted by crosses and the solid circles denote screened E_1 sites. (iii) Same for particles larger than $a\sqrt{2}$, defining E_3. E_2 can be reached by particles slightly larger than a. After Grossman and Aharony (1986, 1987).

the walker is forced to move backwards only if all other alternatives are vacant. As the procedure is continued, the walker reaches the other endpoint. Figure 33(a) shows an example of a cluster, with the numbers indicating the steps of the left-turning walker. Note that after step 17 the walker finds itself again at site number 10. It is thus clear that all the sites between 10 and 17, and all the sites connected to them, are external dangling sites. The next step back to 9 verifies that 10 is also a dangling site. The same is true for sites 24 to 26.

The procedure is now repeated with a right-turning walker. The steps of this walker are denoted by letters on Fig. 33(a). The sites visited by both walkers, i.e. 1, 6, 7, 8, 9, 18, 19, 20, 21, 29, 30 and 31 in Fig. 33(a), are singly connected. The algorithm may thus be used to study quantitative details of the singly connected bonds, the blobs, etc.

The *hull* of the cluster is identified as the total number of sites visited by either walker. Heuristic arguments (see next section), Monte Carlo simulations (Ziff *et al.*, 1986) and exact calculations (Saleur and Duplantier, 1987) all show that the hull is a fractal, with a fractal dimensionality equal to $D_h = 7/4$. Among other physical applications, the hull is the path followed by an electron on the cluster under a very strong magnetic field perpendicular to it; the electron then attempts to maximize the area surrounded by its closed path (Mehr and Aharony, 1988).

Is the hull related to the external perimeter searched in the beginning of this section? Returning to Fig. 33(a), look at sites number 16 and 26. If the diffusing particle is larger than the next nearest neighbour distance on the lattice (i.e. $a\sqrt{2}$, where a is the lattice constant, assuming the cluster sites are point-like), then it will not be able to diffuse through the space between these two sites. As a result, sites numbers 17, 9, 21 and 22 will not be accessible from the outside. The definition of the external perimeter thus depends on the size of the probing particles! The hull (more precisely, its vacant external neighbours) will represent the accessible external perimeter provided that size is smaller than $a\sqrt{2}$. Another way to phrase this is to say that the external sites, which neighbour the hull, are connected to infinity via either nearest or next nearest neighbour unoccupied sites. The diffusing particles should thus be allowed to hop to either nearest or next nearest neighbour vacant sites.

If the size of the diffusing particle is larger than $2a$, then it will also not be able to enter into the gate between h and j in Fig. 33(a). This will then also exclude sites 20 and i from the external perimeter. It is thus clear that smaller and smaller subsets of the hull may serve as adsorbing or reacting sites, depending on the size of the probing particles.

Instead of looking at cluster sites, we can also look at vacant external sites. If each occupied site represents a full square, then the 'gate' between sites 16 and 26 will always be blocked. Figure 33(b) now defines three types of accessible external perimeters, and Fig. 34 shows these perimeters on a large cluster. Looking at Fig. 34, we might think that H, E_1, E_2 and E_3 have decreasing fractal dimensions. However, the situation turns out not to be so complicated: E_1, E_2 and E_3 all have the same asymptotic fractal dimension, which is very close to $D_e = 4/3$. This happens to be equal to the fractal dimension of two-dimensional self-avoiding walks (see Appendix B), and there exist heuristic arguments which relate the two to each other.

On the triangular lattice, E_1 turns out to scale with D_h, while E_2 and E_3 scale with D_e.

In three dimensions, as noted in Section 3.1, practically every site on the cluster is connected via vacant sites to infinity. Therefore, the external perimeter is proportional to the mass s of the cluster, as is the total perimeter including the internal holes already in two dimensions.

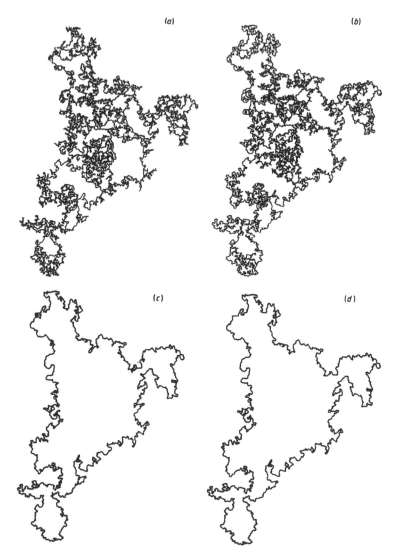

Fig. 34.   The hull and the accessible perimeters of a large cluster on the square lattice at p_c: (a) the hull (10 734) sites; (b) the perimeter E_1 (10 932 sites); (c) the perimeter E_2 (3560 sites); (d) the perimeter E_3 (3284 sites). From Grossman and Aharony (1987).

6.5. DIFFUSION FRONTS

The percolation clusters discussed so far represented the randomness of the *medium*, e.g. the pore structure of a given rock. In the last two sections of this chapter and in the next chapter we describe *dynamic algorithms*, under which the clusters grow and change with time. The first of these, concerning the diffusion of particles from a line (or plane) source, was studied in great detail by Sapoval *et al.* The example of diffusion on the square lattice is shown in Fig. 35. Particles come from a source at the lefthand edge of the illustration, where the particle concentration is kept equal to unity. Any one of the particles is then allowed to hop to one of its four neighbouring sites, provided that it is empty. Since the displacements in the x and y directions (perpendicular and parallel to the source) are statistically independent, the mean square distance of the particles from the source after time t is given by Eq. (106a),

$$\langle x^2 \rangle = \mathscr{D}t$$

Solving Eq. (104b) with the boundary condition $P = 1$ at $x = 0$ now yields

$$P(x, t) = 1 - \frac{2}{\sqrt{2\pi \mathscr{D}t}} \int_0^x dx' \, e^{-(x')^2/(2\mathscr{D}t)}$$

$$= 1 - \frac{2}{\sqrt{2\pi}} \int_0^{x/\sqrt{\mathscr{D}t}} du \, e^{-u^2/2} \tag{129}$$

$P(x, t)$ is the probability that a site at distance x from the source is occupied by a particle after time t. By symmetry, P is independent of y. In contrast with 'normal' percolation, where the occupation probability is constant, we now have an occupation probability $P(x, t)$ which varies with both x and t.

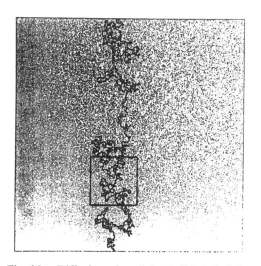 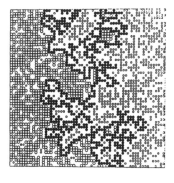

Fig. 35. Diffusion of particles (solid circles) from a source at the lefthand edge of the paper lattice of sites (open squares). The particles (solid squares) that are both connected to the source and neighbours to the righthand edge constitute the *hull*, or the diffusion front. The righthand figure is a blow-up of the region marked by a square on the lefthand figure. Here the hull is shown by solid circles, sites connected to the source as large circles, and the remaining particles as small circles. After Feder (1988), cited in Chapter 1.

At fixed t, $P(x, t)$ decreases monotonically from 1 (at $x = 0$) to zero (for $x \gg \sqrt{\mathcal{D}t}$). Near the source, where $P(x, t)$ is close to 1, the occupied sites form a rather compact and homogeneous cluster. For large x, $P(x, t)$ is smaller than the square lattice percolation threshold p_c, and there arise only small isolated clusters (formed of individualistic particles which chose to diffuse forward and escape from their dense community). We expect the cluster of sites which are connected to the source to end, and the largest finite clusters to appear (on the average), at a distance x_c where

$$P(x_c, t) = p_c$$

From Eq. (129), it is clear that x_c is proportional to $\sqrt{\mathcal{D}t}$, i.e. to $\sqrt{\langle x^2 \rangle}$.

For a variety of reasons (similar to those discussed in the last section) we are interested in the diffusion front, i.e. the hull of the cluster which is connected to the source. As stated above, we expect this front to move forward with time, so that the average x-coordinate of its sites is equal to x_c. Near the percolation threshold p_c, the characteristic linear size of a finite cluster (which may become part of the hull by a movement of a few particles) is given by the percolation correlation length, ξ. We thus identify ξ as a *width of the hull*. Since P depends on x, ξ is never infinite. To calculate it, we must solve the self-consistent equation

$$\xi \propto |P(x_c + \xi, t) - p_c|^{-\nu}$$

Expanding $P(x, t) - p_c = dP/dx|_{x_c}(x - x_c)$, and noting from Eq. (129) that at $x = x_c \propto \sqrt{\mathcal{D}t}$ one has $dP/dx|_{x_c} \propto 1/\sqrt{\mathcal{D}t} \propto 1/x_c$, we find that

$$\xi \propto \left| \xi \frac{dP}{dx} \right|_{x_c} \right|^{-\nu} \propto (\xi/x_c)^{-\nu}$$

i.e.

$$\xi \propto x_c^{\nu/(1+\nu)} \tag{130}$$

Indeed, Sapoval *et al.* confirmed this relation numerically, with $\nu/(1 + \nu) = 4/7 = 0 \cdot 57$. They also noted that the above procedure, of finding the average location of the hull, is a very accurate method to determine p_c in two dimensions.

Equation (130) shows that as time progresses the width of the front ξ becomes smaller and smaller relative to the cluster size x_c. If we perform a renormalization group transformation, grouping sites into cells of size $b \times b$, the width will rescale by a factor $b^{\nu/(1+\nu)}$, whereas the size in the y direction will rescale by a factor b. Such curves, which rescale differently in the longitudinal and transverse directions, are called *self-affine*, instead of self-similar.

One way to measure the fractal dimension of a self-avoiding walk, like the front, is to measure its length using rulers of different sizes δ. For a self-similar curve, the number of such rulers increases with decreasing δ as δ^{-D}, and therefore the total measured length grows as δ^{1-D}. (This is the 'standard' way to measure e.g. the length of the coastline of Norway, and to realize that the result depends on the size of the legs of the walker who tries to measure it by counting steps. See the books by Mandelbrot and by Feder for more details.) On our self-affine curve, the projections of δ on the y and x axes scale as Δy and $\Delta y^{\nu/(1+\nu)}$ respectively, and thus

$$\delta = \sqrt{A(\Delta y)^2 + B(\Delta y)^{2\nu/(1+\nu)}}$$

If the system's size in the y direction is L, then the number of such rulers is of order $L/\Delta y$, and therefore the total measured length is $\delta L/\Delta y$. For very large Δy, the above expression for δ yields $\Delta y \propto \delta$, hence the curve is linear, with length proportional to L. For small Δy, δ is dominated by the second term, $\delta \propto \Delta y^{\nu/(1+\nu)}$, and the length of the curve becomes $\delta L/\delta^{(1+\nu)/\nu}$. Comparing with the fractal result δ^{1-D}, we identify a local fractal dimension

$$D_h = \frac{1+\nu}{\nu}$$

Setting $\nu = 4/3$ reproduces the result $D_h = 7/4$, quoted in Section 6.4. A related heuristic argument led Sapoval *et al.* to conjecture that this result is exact, as indeed later proved by Saleur and Duplantier (1987). Generally, we expect a crossover from a length proportional to L^{D_h}, for $L < \xi$, to one proportional to L, for $L > \xi$.

As for percolation hulls, one can also measure the various accessible perimeters of the diffusion front, for large probing particles. In two dimensions, Sapoval *et al.* indeed find that on scales $L \ll \xi$ they have fractal dimensions $D_e = 4/3$, as for 'normal' percolation.

Diffusion fronts were also studied in three dimensions (Rosso *et al.*, 1986). For site percolation on the cubic lattice, $p_c = 0.3116$. An empty site is considered connected to another empty site if it is on one of the 26 neighbouring sites in a cube of side $3a$, centred on it. The percolation threshold for this connectivity is about 0.097. Thus, for $0.3116 < p < 0.903$ there is connectivity of both occupied and empty sites, with the 'infinite' clusters of both kinds interpenetrating each other. As a result, a finite fraction of the occupied sites belong to the diffusion front, and its fractal dimension is equal to the Euclidean dimension $d = 3$.

6.6. INVASION PERCOLATION

Invasion percolation is a dynamic percolation process which imitates the displacement of one fluid by another in a porous medium. When water is injected very slowly into a porous medium filled with oil, the capillary forces dominate the viscous forces, and the dynamics is determined by the local pore radius r. Capillary forces are strongest at the narrowest pore necks. It is consistent with experimental observations to represent the displacement as a series of discrete jumps in which at each time step the water displaces oil from the smallest available pore.

Wilkinson and Willemsen (1983) simulated the model on a regular lattice. Sites and bonds represented pores and throats and were assigned random 'radii'. For convenience, one assumes that the easily invaded throats are invaded instantaneously, and one assigns random numbers r in the range $[0, 1]$ representing the pore sizes, to the sites. *Growth sites* are identified as the sites that belong to the 'defending' fluid and are neighbours to the invading fluid. At every time step the invading fluid is advanced to the growth site that has the lowest random number r.

The invading fluid may trap regions of the defending fluid. As the invader advances it is possible for it to completely surround regions of the defending fluid, i.e. completely disconnect finite clusters of the defending fluid from the exit sites of the sample. This is one origin of the phenomenon of 'residual oil', a great economic problem to the oil industry. Since oil is incompressible, Wilkinson and Willemsen introduced the rule that the water cannot invade trapped regions of oil. This rule is implemented by removal of growth sites in the regions completely surrounded by the invading fluid from the list of growth sites. Figure 36 shows the results of a simulation of the invasion process. A colour version of this figure appears on the cover of this book. There, each colour indicates sites added with a time interval $\Delta t = 2121$. At the time

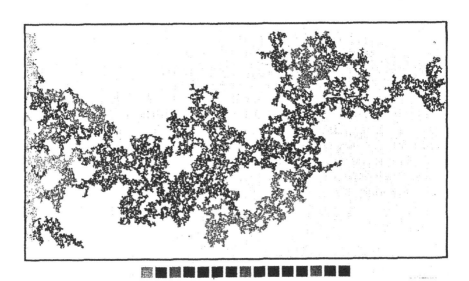

Fig. 36. (See cover picture for colour.) Invasion percolation with trapping in a 300×600 lattice. The invader (coloured) enters from sites on the lefthand edge and the defender (white) escapes through the righthand edge. At breakthrough the invader first reaches the righthand edge and has invaded 31 802 sites. Different colours (left to right on colour scale) indicate sites added within successive intervals $\Delta t = 2121$. From Furuberg *et al.* (1988).

of breakthrough, the number of sites that belong to the central $L \times L$ part of an $L \times 2L$ lattice, with injection from one side, scales as L^D, with $D = 1 \cdot 82$.

It is interesting to note that once the invader reaches the other end of the sample, there is no reason to invade more pores: the invader now has an open path from one end to the other. Thus, the algorithm works so as to fill just the minimum necessary sites for connecting the two ends, and the spanning cluster is a fractal, i.e. it is at its 'critical point'. Processes which build up so as to stay at their critical points have recently been called 'self-organized critical' (Bak and Chen, 1989). Understanding such processes may help us understand why so many natural phenomena are fractals.

The results described above should be contrasted with the ordinary percolation process, for which the cluster is found by occupation of all available sites with random numbers $r \leqslant p$ that are connected to the seed (or to the source), and p is a pre-chosen occupation probability. There are two main differences between ordinary and invasion percolation. First, invasion percolation will always span the region between the injection and the extraction sites. There is no analogue to the occupation probability p, and there are no 'invader' finite clusters. Second, invasion percolation is a dynamic process, with a well defined sequence of invaded sites, as can be seen in Fig. 36: after a new pore is invaded, one opens many easy nearby pores and about l^D new sites are invaded in a local region of size l before growth moves to another place (Furuberg *et al.* 1988).

Wilkinson and Willemsen also studied invasion percolation without trapping, which is possibly appropriate if the 'defending' fluid is compressible. At breakthrough in two dimensions they found that the fractal dimension of the invading fluid is $1 \cdot 89$, similar to that of the spanning percolation cluster at p_c. There is now considerable evidence that invasion percolation without trapping is indeed equivalent to ordinary percolation (Dias and Wilkinson, 1986).

For two-dimensional invasion percolation both with and without trapping, the measured hull and accessible external perimeters were found to have the 'normal' percolation fractal dimensions $D_h = 7/4$ and $D_e = 4/3$. This is a simple consequence from the fact that sites on the hull do not feel the trapping, which happens only behind the front.

In three dimensions, trapping practically never happens, since almost every cluster site also belongs to the hull. Thus, both kinds of invasion percolation have at breakthrough a fractal dimension $D = 2 \cdot 5$, equal to that of 'normal' percolation.

FURTHER READING

For a review on dynamics near thermal phase transitions, see Hohenberg, P.C. and Halperin, B.I., *Rev. Mod. Phys.* **49**, 435 (1977).

Many of the topics discussed in this chapter are also reviewed in papers by Voss,

Stinchcombe, Stanley, Aharony, Orbach, Ziman, Guyon and others in *Scaling Phenomena in Disordered Systems*, edited by Pynn, R. and Skjeltorp, A. (New York: Plenum Press, 1985). See also papers by Aharony, Entin-Wohlman, Courtens, Rammal and Harris in *Time Dependent Effects in Disordered Materials*, edited by Pynn, R. and Riste, T. (New York: Plenum Press, 1987).

For a general review on diffusion, see Havlin, S. and Ben-Avraham, D. *Adv. Phys.*, **36**, 695 (1987).

Analytic, scaling, and series results on diffusion
Gefen, Y., Aharony, A. and Alexander, S., *Phys. Rev. Lett.*, **50**, 77 (1983).
Harris, A.B., Meir, Y. and Aharony, A., *Phys. Rev. B*, **36**, 8752 (1987).

Monte Carlo simulations of diffusion
Ben Avraham, D. and Havlin, S., *J. Phys. A*, **15**, L691 (1982).
Pandey, R.B., Stauffer, D., Margolina, A. and Zabotlitzky, J.G., *J. Stat. Phys.*, **34**, 427 (1984).
Roman, H.E., *J. Stat. Phys.*, **59**, 375 (1990) and **64**, 851 (1991).

'Termites'
Leyvraz, F., Adler, J., Aharony, A., Bunde, A., Coniglio, A., Hong, D.C., Stanley, H.E. and Stauffer, D., *J. Phys. A*, **19**, 3683 (1986).

Directed percolation
Duarte, J.A.M.S., *Z. Physik*, **80**, 299 (1990).

Probability distributions, superlocalization, etc.
Aharony, A. and Harris, A.B., *Physica A*, **163**, 38 (1990).
Bunde, A., Havlin, S. and Roman, H.E., *Phys. Rev. A*, **42**, 6274 (1990).

Fractons
Alexander, S. and Orbach, R., *J. Physique (Paris) Lettres*, **43**, L625 (1982).
Rammal, R. and Toulouse, G., *J. Physique (Paris) Lettres*, **44**, L13 (1983).

Quantum percolation
Meir, Y., Aharony, A. and Harris, A.B., *Europhys. Lett.* **10**, 275 (1989).

Hulls and perimeters
Grossman, T. and Aharony, A., *J. Phys. A*, **16**, L745 (1986); **20**, L1193 (1987).
Mehr, R. and Aharony, A., *Phys. Rev. B*, **37**, 6349 (1988).
Saleur, H. and Duplantier, B., *Phys. Rev. Lett.*, **58**, 2325 (1987).
Ziff, R.M., Cummings, P.T. and Stell, G., *J. Phys. A*, **17**, 3009 (1984).

Diffusion fronts
Rosso, M., Gouyet, J.F. and Sapoval, B., *Phys. Rev. Lett.*, **57**, 3195 (1986).
Sapoval, B., Rosso, M. and Gouyet, J., in: *The Fractal Approach to Heterogeneous Chemistry* mentioned after the introduction.

Invasion percolation
Dias, M.M. and Wilkinson, D.J., *J. Phys. A*, **19**, 3131 (1986).
Furuberg, L., Feder, J., Aharony, A. and Jøssang, T., *Phys. Rev. Lett.*, **61**, 2117 (1988).
Wilkinson, D.J. and Willemsen, J., *J. Phys. A*, **16**, 3365 (1983).

Self-organized criticality
Bak, P. and Chen, K., *Physica D*, **38**, 5 (1989).

CHAPTER 7

Application to Thermal Phase Transitions

Criminals are said always to return to the site of their crime, and thus we now return to one of the original motivations for percolation research, the droplet description for thermal critical phenomena. We summarize what percolation theory for cluster numbers tells us about its 'ancestor'. Dilute Ising models combine the thermal with the geometrical approach, and thus both start and finish this chapter.

7.1. STATISTICAL PHYSICS AND THE ISING MODEL

In thermal physics we deal with the effect of a finite *temperature*, so far ignored. If one atom can be in two states, with energies E_1 and E_2, then at an absolute temperature T it can be found in these states with probabilities proportional to $\exp(-E_1/kT)$ and $\exp(-E_2/kT)$, respectively, where $k = 1 \cdot 4 \times 10^{-23}$ joules/kelvin is Boltzmann's constant. Generally, a configuration with energy E can be found with a probability proportional to $\exp(-E/kT)$. The factor of proportionality is such that the sum over all probabilities is unity; thus the thermal probability is

$$p_i = \frac{\exp(-E_i/kT)}{\sum_j \exp(-E_j/kT)}$$

where the sum goes over all possible states i ($i = 1$ and $i = 2$ in the above example) and thus involves all energies. The thermal average $\langle A \rangle$ of some quantity A, having the value A_i in state i, is then

$$\langle A \rangle = \sum_i p_i A_i = \frac{\sum_i A_i \exp(-E_i/kT)}{\sum_i \exp(-E_i/kT)}$$

Averages of interest are the energy $\langle E \rangle$, the specific heat $C_v = d\langle E \rangle/dT$, and many others.

For example, the two energy levels above may correspond to a magnetic dipole in a magnetic field H (often this field is also denoted by B): $E_1 = -H$, $E_2 = +H$ in suitable units for the field. There we assume that the magnetic dipole can only point up or down, parallel or antiparallel to the field H, and we call this dipole associated with an atom its 'spin'. In a ferromagnetic material like iron, neighbouring spins tend to be parallel to each other. A pair of spins then has the 'exchange' energy $-J$, if it is parallel, and $+J$ if it is antiparallel. The total energy E in a field H is then, with spins $S_i = \pm 1$:

$$E = -J \sum_{\langle i, k \rangle} S_i S_k - H \sum_i S_i$$

Here the first sum runs only over nearest neighbour pairs of the lattice and contains each such pair only once, e.g. by $i < k$.

This model is the well-known Ising magnet. It can also be rewritten as a lattice gas model for fluids, where spin up corresponds to an occupied site and spin down to an empty site. Regions with most spins up then represent liquid, regions with most spins down are identified with vapour.

The Curie point of a ferromagnet is that temperature T_c below which for zero field a spontaneous magnetization m_0 appears. Generally, the magnetization m (in suitable units) is the difference between the number of up and down spins. Similarly, below the critical temperature T_c of a fluid, separation of liquid and gas is possible; the density difference between a liquid and its vapour is then the analogue of $2m_0$. The susceptibility χ is the zero-field derivative dm/dH, and its fluid analogue is proportional to the compressibility, since H corresponds to something like the pressure. Near the Curie or critical point we find the critical exponents for $H = 0$:

$$C_v \propto |T - T_c|^{-\alpha}$$

$$m_0 \propto (T_c - T)^{\beta}$$

$$\chi \propto |T - T_c|^{-\gamma}$$

$$\xi \propto |T - T_c|^{-\nu}$$

Here the correlation length ξ indicates the range over which one spin influences appreciably the orientations of other spins.

These critical exponents of the Ising model are known exactly in two dimensions ($\alpha = 0$, $\beta = 1/8$, $\gamma = 7/4$, $\nu = 1$) and numerically in three: ($\alpha = 0 \cdot 11$, $\beta = 0 \cdot 32$, $\gamma = 1 \cdot 24$, $\nu = 0 \cdot 63$). They fulfil the scaling law $2 - \alpha = \gamma + 2\beta = d\nu$ in $d = 2$ and 3 dimensions, just as in percolation. The three-dimensional Ising exponents agree well with those of real fluids.

In the analogy with percolation, we thus identify the spontaneous magnetization with the strength of the infinite cluster, the susceptibility with the mean cluster size, and temperatures T above T_c with concentrations p below p_c. Correlation lengths for magnets correspond to correlation lengths for percolation. Rather generally, the analogue of spontaneous magnetization or strength of the infinite cluster is called the *order parameter* of a phase transition and is zero on one side of that transition; it vanishes with the exponent β if this transition is approached from the other side. Of course, no analogy is perfect: the order parameter in a ferromagnet (the spontaneous magnetization) can show in at least two different directions, whereas the strength P of the infinite cluster has a unique value.

7.2. DILUTE MAGNETS AT LOW TEMPERATURES

Is this analogy with percolation accidental? (Would we write about it if it were? Hardly!) Imagine that only a fraction p of all lattice sites is occupied by spins and the remaining fraction $1 - p$ is left non-magnetic. These spins are distributed randomly, as in percolation. This model is called the site-diluted quenched Ising model. ('Annealed', instead of 'quenched', systems are systems where the spins and the non-magnetic atoms can interchange places according to some thermal equilibrium; we ignore 'annealed' dilution here.)

At low temperatures ($kT \ll J$ but $H \propto T$) spins within one percolation cluster will be parallel to each other in equilibrium; the probability to flip spins in them involves powers of $\exp(-2J/kT)$ since $2J$ is the energy to 'break' one bond. On the other hand, different clusters have different orientations and do not influence each other. Thus each finite percolation cluster of s sites acts as if it were one super-spin with an s times larger magnetic moment and thus an energy $\pm sH$ in a field H. Thus its probability to point in the direction of H is

$$\frac{e^{sH/kT}}{e^{sH/kT} + e^{-sH/kT}}$$

whereas the probability to point in the opposite direction is

$$\frac{e^{-sH/kT}}{e^{sH/kT} + e^{-sH/kT}}$$

With the hyperbolic tangent $\tanh(x) = (e^x - e^{-x})/(e^x + e^{-x})$ the difference between these two probabilities, multiplied with the size s, is thus the magnetization per cluster:

$$m_{\text{cluster}} = s \tanh(sH/kT)$$

The infinite cluster if present contributes $\pm P$ to the magnetization m, depending on the orientation. Thus the total magnetization per lattice site is

$$m = \pm P + \sum_{\text{cluster}} m_{\text{cluster}} = \pm P + \sum_s sn_s \tanh(sH/kT)$$

For $H \to 0$, only the infinite cluster remains: $m_0 = \pm P$. Expanding $\tanh(x) = x - O(x^3)$ for small fields H, we find the susceptibility χ (equal to the derivative dm/dH at zero field H) to be essentially the mean cluster size S:

$$\chi = \sum_s \frac{s^2 n_s}{kT} \propto S$$

Thus, if at low temperatures the concentration p of spins is varied to approach the percolation threshold, then $m_0 \propto (p - p_c)^\beta$, $\chi \propto |p - p_c|^{-\gamma}$ with the percolation exponents β, γ and not the undiluted Ising exponents. Thus, in this case we find an exact correspondence of the infinite cluster to the spontaneous magnetization, of the mean cluster size to the susceptibility, and of the percolation threshold to the transition for ferromagnetism.

While clusters are well defined to describe geometrically the magnetic behaviour for these low-temperature dilute Ising models, their proper definition for pure Ising models at $T \to T_c$ is the success story to be told in the next sections. Then we will return to the dilute Ising case.

7.3. HISTORY OF DROPLET DESCRIPTIONS FOR FLUIDS

Now we review the recent progress in the decade-long attempts to describe thermal phase transitions such as the Curie point of ferromagnets or the formation of a liquid out of vapour through a cluster or droplet model. These droplets are a modification of the percolation clusters discussed so far and were alluded to in Section 2.5. Note that we return here to the non-dilute case. The aim of these efforts as now described in Sections 7.3–7.6 is to have a geometric interpretation of the physical phase transitions: can we describe the boiling of water and other liquid–gas transitions through a percolation picture, with clusters propagating the correlations, as was possible for the dilute magnet at low temperatures?

When it rains, tiny water droplets are formed out of a supersaturated vapour. The relative humidity tells us whether a vapour is supersaturated and is the ratio of the actual water vapour pressure to the saturated vapour pressure; at saturation, i.e. at 100 per cent relative humidity, bulk liquid and vapour are in equilibrium. To form the first droplet out of the vapour, an energy barrier due to the surface tension between liquid and vapour must be overcome through thermal activation; once small droplets (larger than some minimum size) are formed, they can grow more easily and fall down to earth. (Similarly, the main problem with your bank account is to get the first million pounds; from then the account grows much more easily, as you surely remember.) This process is called *nucleation*.

Nucleation theory since 1925 usually assumes that the surface area of a small droplet varies in three dimensions as the square of the radius, i.e. as $s^{2/3}$ if the droplet contains s water molecules. The rate at which droplets are nucleated is assumed to be proportional to $\exp(-E/kT)$ at temperature T, where $E \propto s^{2/3}$ is the energy barrier due to the formation of a surface between liquid and vapour; the proportionality factor is basically the surface tension.

This classical nucleation theory is, in general, in good agreement with both laboratory and computer experiments. For small supersaturations, this nucleation rate can be so small that one would have to wait for years to see the first droplet (metastable equilibrium). In this case, the presence of small solid particles, serving as nucleation centres, helps the droplet to overcome the nucleation barrier E (as a loan from your bank allows you to become rich more easily, as long as you do not have to pay it back).

Also in a vapour which is undersaturated, i.e. which is in complete and not in metastable equilibrium, tiny droplets can form and decay again. If no such droplets occur at all, then we have the classical ideal gas law: $P = nkT$ for $N = nV$ separate molecules in a volume V under the pressure P. If instead we have a few pairs ($s = 2$), triplets ($s = 3$), and other small clusters containing s molecules each, then their total pressure is

$$P = \sum_s n_s kT$$

with n_s such clusters per unit volume; if only $s = 1$ is important in this sum we go back to the ideal gas. In this approximation we neglect the forces between different clusters and then can simply add the partial pressures produced by the different clusters, just as the air pressure is basically the sum of the nitrogen and oxygen pressures.

Such cluster pictures and more formal cluster expansions have been discussed since about 1940. The Fisher droplet model (see Section 2.5) replaces in the droplet formation (free) energy E_s the contribution proportional to $s^{2/3}$ by one proportional to s^σ and adds $\tau \ln(s)$ to E_s; thus in equilibrium,

$$n_s \propto \exp(-E_s/kT) \propto s^{-\tau} \exp(-\Gamma s^\sigma)$$

with a temperature-dependent microscopic surface tension Γ. We see that the scaling law (33) for percolation cluster numbers is merely a slight modification of this Fisher droplet model; in fact, Eq. (33) was suggested for thermal phase transitions by Binder before it was suggested for percolation clusters, around 1975.

The quantitative test of these ideas was hampered, however, by the fact that no precise definition of a cluster or droplet was available at that time for thermal phenomena. Even today, researchers debate the definition of a hydrogen bond keeping water clusters (ice) together. Thus the Ising model of the preceding section seemed at first a nice description: clusters are groups of neighbouring occupied lattice sites, just as for percolation. Unfortunately that did not work properly in three dimensions. A modified definition worked for saturated vapour but led to difficulties for temperatures above the critical temperature. Only with the inclusion of a ghost spin could that difficulty be removed, and this Swendsen–Wang formulation of 1987 gave a proper description of both equilibrium cluster numbers and nucleation phenomena.

The Swendsen–Wang description not only gave a geometric understanding of thermal transitions through clusters or droplets, but their algorithm also saved computer time. It no longer deals with single molecules but with whole clusters, and thus reaches equilibrium much faster, as we will show. Similarly, theories of solid state physics do not always deal with each atom independently; usually they work with phonons (sound waves) which are nearly independent of each other and thus add up to give, for example, the specific heat of the Debye law, $C_v \propto T^3$. Just as these phonons are the elementary excitations of solids, our clusters or droplets are now thought to be the elementary excitations for fluids (or anisotropic magnets). Addition of the nearly independent contributions from each such elementary excitation gives the property of the whole system. Fortunately, experimental techniques have also developed to such an extent that since 1988 clusters can be made visible in suitably chosen fluids. We now give in greater detail the definition of these clusters and their properties as found by computer simulations.

7.4. DROPLET DEFINITION FOR ISING MODEL IN ZERO FIELD

For clarity we now describe as 'clusters' the groups of neighbouring occupied sites familiar from percolation theory, and as 'droplets' the properly defined elementary excitations which describe the behaviour of the fluid. Such droplets should satisfy the conditions which the percolation clusters do, in particular near the critical point where the ideal gas law fails completely. The dilute Ising model for $T \rightarrow 0$ obeyed them trivially:

1. An infinite droplet is formed for the first time at and only at the critical point.
2. The order parameter of the transition, i.e the density difference between liquid and gas for fluids, and the spontaneous magnetization m_0 for magnets, is related to the size of the infinite droplet.
3. The 'susceptibility' of the order parameter, i.e the compressibility of the fluid, is proportional to the mean cluster size, i.e. to the second moment $\Sigma_s s^2 n_s$.
4. The correlation length, i.e. the spatial extent of the thermal fluctuations, is proportional to the radius of a typical cluster.

The simple definition of a cluster as a group of neighbouring parallel spins turns out to be wrong: in the simple cubic lattice the up spins already percolate at a temperature several per cent below the critical temperature, at a point when less than 1/4 of all spins point up. This result is somewhat plausible from the fact mentioned earlier (see, e.g., the Bethe lattice result) that the percolation threshold goes to zero if the number of neighbours goes to infinity. Thus, provided that the number of neighbours is sufficiently large, the percolation temperature is allowed to be far below the Curie temperature, since at the latter half the spins always point up in zero field.

This error was corrected by the Coniglio–Klein definition, based on a theorem of Kasteleyn and Fortuin: two up spins are regarded as part of the same droplet only if they are connected by an additional bond. These additional bonds are distributed randomly in the lattice between parallel neighbours, with a bond probability $\pi = 1 - \exp(-2J/kT)$. These additional bonds do not enter the interaction energy, they just serve to define droplets appropriately. (The analogous definition was later applied also to down spins.) We thus have here a random bond percolation superimposed on the Ising-correlated site lattice of occupied and empty places. With this definition the Curie temperature, i.e. the critical point, agrees with the percolation threshold, and also the three other criteria above are fulfilled in zero field H; the spontaneous magnetization at zero field is given by the size of the infinite droplet. Simulations also showed the typical droplet radius to agree with the correlation length; at the Curie point these droplets are fractal with a radius increasing as a power of s and a fractal dimension $2 \cdot 5$ on the cubic lattice. Theoretically we expect again $D = d - \beta/\nu$ with, however, the Ising analogues of β, ν, which differ slightly from percolation: $D = 2 \cdot 53$ instead of $2 \cdot 48$. Thus in zero field everything seems fine, and the reader not interested in recent research can skip the next two sections.

7.5. THE TROUBLE WITH KERTÉSZ

As pointed out by Kertész, a problem arises in a magnetic field above the critical temperature. Experimentally and theoretically we know that the equation of state (density as a function of pressure and temperature, or magnetization m as a function of field and temperature) does not have any divergences in any of its derivatives (with respect to H or T) as soon as the field H is non-zero. The Coniglio–Klein droplets, on the other hand, have a whole percolation transition line extending from the Curie point ($H = 0$) to infinite fields; if we cool down the system at constant density then at this line an infinite droplet is formed for the first time even though no spontaneous magnetization appears there. Thus physics and geometry are not in one-to-one correspondence, in contradiction to the first criterion. Also the mean cluster size and the typical cluster radius diverge along this so-called Kertész line, although no divergences appear in the susceptibility or correlation length. The undesired divergences have thus come up again, just like the title character in Hitchcock's movie The Trouble with Harry.

The reason for the existence of this Kertész line is quite plausible: at an infinitely strong up field H, for a finite temperature, all spins point up, and between them the additional bonds are distributed randomly with probability $\pi = 1 - \exp(-2J/kT)$. Thus we now have a normal bond percolation problem, and the condition

$$1 - e^{-2J/kT} = p_c^{\text{bond}} = 0 \cdot 2488 \tag{131}$$

(on the simple cubic lattice) determines a percolation temperature, which is 55 per cent above the Curie temperature of the cubic lattice. From this point at infinite field (unit density), the Kertész line extends smoothly down to the

Curie temperature at zero field (density $1/2$) (Fig. 37). How to get rid of Kertész?

For Ising models in a magnetic field, Swendsen and Wang, following Kasteleyn and Fortuin, introduced a *ghost spin* oriented parallel to the magnetic field. The spins of the lattice parallel to the field and the ghost spin are connected to this ghost spin by additional bonds, distributed randomly with a probability proportional to the field for small H. (In general the probability is $1 - e^{-h}$, $h = 2H/kT$.) For a two-dimensional lattice, we can imagine the ghost spin to sit above the lattice in the third dimension. Even for a very small field, like $h = 0 \cdot 01$, every lattice spin has a small chance (1 per cent) to be connected to the ghost spin. Thus a large lattice of, say, one million sites will have about 10 000 sites, separated from each other by typically 10 lattice constants but connected indirectly via the ghost spin. They constitute an infinite though very loose cluster (in an infinite lattice), like members of a family spread over all continents but still phoning each other via satellite. Thus, exactly as it should be, there is no infinite cluster at zero field; there are no geometrical divergences any more at non-zero field, and the above criteria are always fulfilled. The probability $1 - \exp(-2J/kT)$ ensures that loose structures at high temperatures are not counted as one single droplet; on the other hand, the probability $1 - \exp(-h)$ creates some connectivity when an external field forces the spins to reorient. We leave formal proofs to the experts and merely conclude: the trouble with Kertész is over.

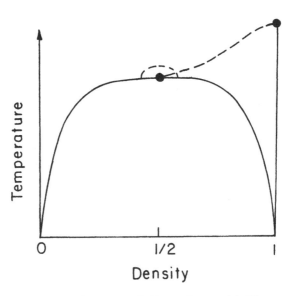

Fig. 37. Schematic phase diagram of a fluid or Ising model. The curve similar to a parabola is the coexistence curve where liquid and vapour coexist. Along the dashed semicircle we can go from the vapour to the liquid phase without any divergences in the density or its derivatives; however, the surface tension for up droplets vanishes along the dashed Kertész line.

7.6. APPLICATIONS

Having found a proper definition, what can we do with it? Among other things we can check the assumptions entering traditional nucleation theories. So we look how the cluster numbers n_s for $T < T_c$ decay as a function of the cluster size s. Computer simulations showed in three dimensions that in zero field the cluster numbers decay as an exponential function of the surface area: $\log n_s \propto -s^{2/3}$. In a field h, it decays as $n_s(H)/n_s(H = 0) \propto \exp(-hs)$ for large hs. These results confirm classical nucleation ideas. Right at the Curie point in zero field the cluster numbers decay as $s^{-2.3}$ in three and as $s^{-2.0}$ in two dimensions, similarly to percolation theory. In this sense these droplets justify the old ideas used to describe the liquid–gas phase transition or its magnetic analogue by the geometric picture of droplets of all sizes. At and above the critical point, the Coniglio–Klein droplets lose their droplet geometry: they no longer have a well-defined surface and instead are fractals similar to *lattice animals*.

There is also a new and surprising prediction arising from this droplet picture. It comes, from the Kertész line which we had hoped to have buried for good, but which is resurfacing again. Along this line the droplet numbers n_s no longer become critical, due to the ghost spin connections which give a factor $\exp(-hs)$ to n_s since each of the s sites of a finite cluster must not be connected to the ghost spin. However, the droplet surface tension Γ is vanishing on the Kertész line, which means that $-\log(n_s)$ varies as $hs + \Gamma s^{2/3} + \cdots$ on the low-temperature side and as $hs + \text{const} \cdot s + \cdots$ on the high-temperature side of this transition line. Possibly this means that Taylor expansions of the free energy as a function of H or T have a different convergence behaviour on the two sides of this line. If correct, it would mean that we have to make more precise the century-old wisdom that 'nothing happens' if we move continuously from the vapour to the liquid phase of a fluid by heating it above the critical temperature. Instead, we say that no singularity in the thermodynamic quantities occurs on this route, but the droplet surface tension vanishes at a *sharp* transition point, namely when crossing the Kertész line (Fig. 37).

Of much greater practical importance is the computer time saved by the Swendsen–Wang algorithm. This method first determines the droplet structure of the spin configuration, and then flips each droplet with probability 1/2. (To get a stable sign of the spontaneous magnetization, or to get proper nucleation events, one may keep the orientation of the largest droplet or ghost spin droplet fixed.) In this way, the system relaxes much faster into equilibrium than with the traditional Metropolis technique of flipping one Ising spin at a time. Right at the Curie point, the relaxation time for a system of linear dimension L varies roughly as L^z. For the traditional single-spin-flip technique z is about 2, whereas for the cluster-flip technique z is much smaller; in two dimensions it could be even zero, corresponding to a $\log(L)$ variation of the relaxation time. If we look at the relaxation of the cluster numbers $n_s(t)$

in the square-lattice Ising model, we find s-dependent relaxation times again varying with $\log(s)$ or some small power of s, if Swendsen–Wang cluster flipping is applied. Thus Ising ferromagnets can now be simulated much better at the Curie point, provided we are not interested in the kinetics of the single-spin-flip method. For example, conflicting theories for two-dimensional randomly diluted Ising models were successfully checked with this method. If similar tricks could be applied to the *lattice gauge* simulations of elementary particle physicists, its implications on computer-time budgets and special-purpose computers would be enormous and could constitute the most important economical application of percolation theory.

7.7. DILUTE MAGNETS AT FINITE TEMPERATURES

We now return to the case where a random fraction p of all lattice sites is occupied by spins whereas the rest is not magnetic. Clusters are now again the usual groups of neighbouring occupied sites.

At finite temperatures, some spins within the same cluster will no longer be parallel to each other. An isolated pair, for example, will be antiparallel with a probability proportional to $1 - \pi = e^{-2J/kT}$. In order still to have a spontaneous magnetization, we now need a higher concentration p, i.e. the ferromagnetic threshold shifts upward by an amount proportional to $1 - \pi$. Finally, for $p = 1$ we recover the pure Ising model; there the Curie temperature T_c is the upper boundary of the region with a spontaneous magnetization: $J/kT_c = \ln(1 + \sqrt{2})/2 = 0 \cdot 44061$ and $J/kT_c = 0 \cdot 22165$ on square and simple cubic lattices for $p = 1$. Figure 38 shows schematically this 'phase diagram'. Only the ferromagnetic region has a spontaneous magnetization.

While for $T \to 0$ we find, as a function of $p - p_c$, percolation exponents for m_0 and χ, at $p = 1$ we find, as a function of $T_c - T$, Ising exponents.

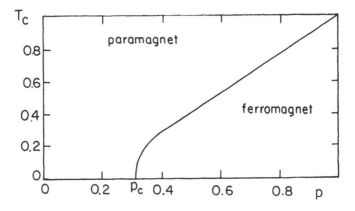

Fig. 38.   Schematic phase diagram of the dilute Ising model on a simple cubic lattice.

In the whole intermediate region, for $p_c(T=0) < p < 1$, a third type of behaviour has been found in two and (less reliably) in three dimensions.

At zero temperature, all the spins on a percolation cluster are parallel to each other. As temperature is slightly raised, some spins will flip, and we shall have Coniglio–Klein thermal droplets superimposed on the quenched geometrical clusters. The Coniglio–Klein bond-preserving probability π is thus very similar to that discussed in Section 4.5. Indeed, the singly connected bonds will also be the first to break at finite temperature. At the percolation threshold $p = p_c$, the bond-breaking probability $1 - \pi$ must obey the same recursion relation as in Eq. (70), hence

$$e^{-2J'/kT} = b^{1/\nu}e^{-2J/kT} + O(e^{-4J/kT}) \qquad (132)$$

Recalling that $\xi' = \xi/b$, this allows us to identify a *thermal correlation length*,

$$\xi_T \propto e^{-2J\nu/kT} \qquad (p = p_c) \qquad (133)$$

For finite $p - p_c$, we end up with the two competing lengths, the geometric length $\propto |p - p_c|^{-\nu}$ and the thermal length ξ_T. Magnetic correlations are destroyed by the geometry if $\xi \ll \xi_T$, and by thermal fluctuations if $\xi \gg \xi_T$.

At finite temperature, or $\pi < 1$, we expect a phase transition at a finite temperature $T_c(p)$. From scaling, we expect the magnetization to behave near p_c, and thus at rather low temperatures, as

$$m = (p - p_c)^\beta m_1(\xi_T/\xi) = (p - p_c)^\beta m_2 \left(\frac{e^{-2J/kT}}{p - p_c} \right) \qquad (134)$$

For positive $p - p_c$, this magnetization m can vanish only if the functions m_1, m_2 vanish. Making the usual power-law assumption $m_2(x) \propto (x - x_c)^{\beta_T}$, we find

$$m \propto (p - p_c)^{\beta - \beta_T}(T_c - T)^{\beta_T} \qquad (135)$$

and $T_c(p)$ is given by

$$\exp(-2J/kT_c) = \text{const} \times (p - p_c) \qquad (136)$$

An alternative way to derive the recursion relation (132) is as follows: consider the two spins at the two opposite edges of the renormalization cell of length b. They are connected via a link containing singly connected bonds and blobs. If we replace this whole link by an effective bond, with an exchange energy J', then the probability p_1 to have the two spins antiparallel is smaller by a factor $y = \exp(-2J'/kT)$ than the probability p_2 of having them parallel: $p_1 = y/(1 + y)$, $p_2 = 1/(1 + y)$. The same rule applies, with J replacing J', for each original bond of the cluster. Each bond on which two spins are antiparallel costs a factor $\exp(-2J/kT)$ in probability. Thus the most probable way to have the two end spins antiparallel is by selecting one of the singly connected bonds and then having all spins on one side of this bond oriented up, and all the remaining spins on the other side of this bond oriented down. Since there are $b^{1/\nu}$ such bonds to select, the total probability for antiparallel end spins is $b^{1/\nu}e^{-2J/kT}$ for small $\exp(-2J/kT)$, apart from corrections of order $\exp(-4J/kT)$ from the denominator. (If we want to have the spins on two ends of a blob antiparallel, we must break at least two bonds which connect these spins in parallel. The corresponding probability again is of order $\exp(-4J/kT)$ and thus also negligible.) This leaves us with Eq. (132).

Magnetic systems are often described by the Heisenberg model, in which the local magnetic moments are three-component vectors, S_i, which are free to rotate in space. The 'exchange' energy of two neighbouring spins is now the scalar product $-JS_i \cdot S_j$. As mentioned in Section 6.3, the basic excitations there are spin waves, and not droplets. If we look at the link described in the previous paragraph, then we shall not have a sharp domain wall between up and down spins. Rather, the spins form a 'Bloch wall' in which they rotate gradually. The mapping alluded to in Section 6.3 can be used to show that in this case, at low temperatures,

$$\frac{J'}{kT} = \frac{b^{\tilde{\zeta}_R} J}{kT}$$

and thus the thermal correlation length behaves as $\xi_T \propto T^{-1/\tilde{\zeta}_R}$. This yields a phase diagram similar to Fig. 38, but with $T_c \propto (p - p_c)^{\tilde{\zeta}_R \nu}$.

7.8. SPIN GLASSES

Spin glasses are much nastier cases of disordered magnets. Now the exchange interaction J_{ik} between two neighbouring spins i and k can be either ferromagnetic (positive) or antiferromagnetic (negative). If a ring of four bonds connecting neighbouring spins has, say, three positive and one negative sign, then these spins are 'frustrated', i.e. they must violate at least one bond whatever their orientation. The scientists working on spin glasses have also been frustrated, since they did not even know the spin orientations at zero temperature, i.e. the ground state with the lowest energy. Frustration tends to create many different states with the same energy.

The situation is not trivial even if we take a dilute ferromagnet, as described earlier in this chapter, and replace a small fraction of its bonds by antiferromagnetic ones with negative J. Similarly to the thermal fluctuations, this destroys the ferromagnetic correlations, and forces an increase in the ferromagnetic percolation threshold at zero temperature, and a decrease of the magnetic ordering temperature above that threshold. In a sense frustration allows blobs (like the frustrated ring of four bonds mentioned above) to break the transmission of order at zero temperature as easily as do the singly connected bonds mentioned before and this increases the ferromagnetic threshold.

The dilution of an antiferromagnet with ferromagnetic bonds seems to be an essential feature in many of the much-investigated high-temperature superconductors.

FURTHER READING

For statistical physics of phase transitions, see the books of Stanley, Domb and Green and of Domb and Lebowitz cited in Chapter 1.

Alexandrowicz, Z., *Physica A*, **160**, 310 (1989); **167**, 322 (1990) [*dynamics*].

Binder, K. and Young, A.P., *Rev. Mod. Phys.*, **58**, 801 (1986) [*spin glasses*].

Chowdhury, D., *Spin Glasses and Other Frustrated Systems* (Singapore: World Scientific, 1986).

Heermann, D.W. and Burkitt, A.N., *Physica A*, **162**, 210 (1990) [*dynamics*].

Kertész, J., *Physica A*, **161**, 58 (1989) and **175**, 222 (1991) [*transition line*].

Miranda, E.N., *Physica A*, **175**, 229 and 235; **177**, 381 (1991) [*dynamics*].

Wang, J.S. *et al.*, *Physica A*, **167**, 565 (1990) [*review*]; **161**, 249 (1989) [*droplet numbers and history*]; **164**, 221 (1990) [*dilute Ising model*]; **164**, 240 (1990) [*dynamics*].

CHAPTER 8
Summary

Once your nightmares are over about termites digging tunnels through your brain, or $1 \cdot 56$-dimensional animals burning in forest fires, you may wish to reflect on whether you have learned anything from this book.

Our first aim was to show that percolation is an active field of research. Many of the results presented here were not known at the time most of the prospective readers of this book were born. In general, the knowledge was formed in the order in which the chapters are printed. Thus the kinetic aspect contains, as its reading list suggests, the more recent ideas. A theorem by Dyson tells us that a publication coming out at time t and taking into account research up to time $t - t_0$ will be outdated at time $t + t_0$. If you prepare any seminar talk about some part of this book, you should therefore try to find out what has happened in research more recently, for example by consulting the Science Citation Index after locating a relevant publication from the reading list in this book. In short, theoretical physics in general and percolation theory in particular is a human enterprise and not the fixed body of knowledge which it often appears to be when presented in formal courses within your curriculum.

In addition we hope you have learned what a phase transition is. Admittedly, percolation has a somewhat unusual phase transition since no temperature is involved. Nevertheless, many functions or their derivatives, as a function of a continuously varying parameter p, diverge or vanish at one sharply defined point, the percolation threshold $p = p_c$. In percolation theory these functions are purely geometric properties; at most other phase transitions one deals with thermal properties like specific heats as function of temperature, etc. But both cases are similar in that the important functions, or their derivatives, are not continuous at the critical point.

The similarity between thermal phase transitions and the percolation threshold becomes even clearer when we look at the scaling laws governing the leading asymptotic behaviour very close to the critical point. Many functions $G(x, y)$ depend on two variables x and y which both vanish right at the critical point. Then, in Eqs. (33), (52), (55), (73), (107) and (115) we have seen

150

six different examples of the same principle:

$$G(x, y) = x^{-A} g(y/x^B)$$

is the scaling assumption for small x and small y. For example, y may correspond to the distance from the critical point, and $1/x$ to the cluster size. Scaling theory does not predict the value of the critical exponents A and B or the precise form of the scaling function g; but with this scaling assumption we can calculate the critical exponents of many other quantities and relate them to A and B. In general, for simple problems two independent exponents like A and B here are sufficient to determine the other exponents. No clear answer could be given, for percolation as well as for thermal phase transitions, as to whether or not the additional exponent entering into dynamical scaling is related to the static exponents. For thermal phase transitions that depends on the system studied. To calculate the numerical value of any of these exponents we have to go beyond scaling theory, since then relations between exponents are not enough. Renormalization group techniques have been developed into a powerful tool for estimating exponents numerically. In a few cases, mostly in two dimensions, we (believe we) know even the exact values.

Right at the critical point, one of the arguments in the function $G(x, y)$ above may be zero; for example $y = 0$ if $y = p - p_c$ or $y = T - T_c$. Then most quantities vary asymptotically with a simple power law, like $G \propto x^{-A}$ for small x. If x happens to be an inverse length one may, under certain conditions of self-similarity, call A the fractal dimension for the quantity G, particularly if G can be identified with the mass of an object. In the largest cluster right at the percolation threshold of an $L \times L \times L$ simple cubic lattice, the number of sites increases asymptotically as $L^{2 \cdot 5}$, which makes it a $2 \cdot 5$-dimensional fractal. We see that in this sense the concept of fractals is contained in the concept of scaling near phase transitions; but the fractal concept can also be applied to power laws where no phase transition occurs, like the lattice animals and their radii discussed after Eq. (50) and the various subsets of the percolation cluster, discussed in Section 5.2.

There is no need to understand thermal physics before one studies percolation. One does not need to know classical or quantum mechanics, or statistical physics, to understand percolation. Only geometry and probability, and for conductivities some elementary concepts of electricity, were required in this book. On the other hand, the knowledge from our simple percolation can be helpful in understanding better the behaviour of the more complicated thermal phase transitions.

Finally we hope that you have learned that percolation is a good introduction to the field of computer simulations. The largest systems ever simulated (10^{12} sites) seem to belong to percolation theory. Small systems can be simulated on simple computers and produce nice pictures like Fig. 2. Percolation theory therefore can be and has been used as a way of learning computer usage. For the expert the Appendix A will give somewhat more information, since the basic difficulty of computer experimentation is the interpretation of data, not their production.

We call these computer simulations 'experiments' since in this book we avoided contact with real experiments in the laboratory or in Nature. Such real experiments, of course, have some difficulties. Usually the lattice is not completely periodic but has some defects, some impurities cannot be avoided, the distribution of sites is not completely random, and so on. Therefore this book has tried to introduce the reader to percolation theory, not to percolation experiment. (We saw, however, that computer experiments have difficulties too: the systems are quite small, and for very large systems the random numbers might not be random enough.) These deviations make it in general quite difficult to get an exact correspondence between the percolation model and some real material. However, the principle of universality suggests that the critical exponents found in percolation theory agree with the critical exponents which careful laboratory experiments should give. If nearly all three-dimensional percolation models have the same critical exponents, then these exponents should not depend on such minor problems as lattice defects. This universality is one of the reasons why critical exponents were emphasized so much in this book. Indeed, certain metal–insulator films turned out to mimic excellently the behaviour found in computer simulations of percolation models, as reviewed by Deutscher, Kapitulnik, and Rapaport in the book on *Percolation Structures and Processes* mentioned at the end of Chapter 1. In the same book, Zallen gives a beautiful overview of the many different applications one might find for percolation.

APPENDIX A
Numerical Techniques

In this appendix we discuss how one can estimate asymptotic quantities from exact 'series' data or from Monte Carlo simulations, and we explain how a computer can count clusters in very large lattices. Estimates of asymptotic properties are usually quite accurate. Estimates of systematic errors, on the other hand, are similar to weather prediction: one can never be sure of being right if one has predicted the behaviour at infinity for a quantity known only for finite intervals. Beginners in particular tend to underestimate the size of the systematic errors involved in their extrapolations.

A.1. ANALYSIS OF EXACT DATA

We mentioned in Section 2.7 two methods of series analysis: Ratio and Padé approximation. To apply these methods, we need to know from exact enumeration the coefficients of a power series in the concentration p or in some other suitable variable. Such methods are also important for thermal critical phenomena. In percolation, we often have data on cluster numbers n_s, cluster radii R_s, numbers of lattice animals g_s, or other similar quantities which we want to analyse directly and which are not coefficients of a series.

In this case we may simply introduce the so-called generating function

$$G(\lambda) = \Sigma_s g_s \lambda^s \qquad (A.1)$$

where g_s stands for the quantity we are interested in. If asymptotically for large s the numbers g_s vary as

$$g_s \propto s^{-\theta} (\text{const})^s \qquad (A.2a)$$

as is the case with many of these quantities, then we can evaluate G for λ slightly below 1/const by replacing the sum in Eq. (A.1) with an integral. Straightforward integration, similar to the evaluation of moments of the cluster size distribution in Chapter 2, then gives

$$G \propto \varepsilon^{\theta-1} \qquad (A.2b)$$

for small $\varepsilon = -\ln(\lambda \text{ const}) \propto \text{const}^{-1} - \lambda$. This generating function and its critical exponent $\theta - 1$ can then be analysed with Padé approximations. (If $\theta > 1$ it may be practical to look instead at the generating function of $s^k g_s$

with k chosen to be so large that the generating function diverges at $\lambda = 1/\text{const}$. Equivalently one many look at derivatives of $G(\lambda)$.)

A.2. ANALYSIS OF MONTE CARLO DATA

Monte Carlo data, in contrast to the exact 'series' data mentioned above, have finite statistical errors. Thus except for very high-quality data, it is not recommended to fit straight lines through two consecutive data points. In general the slope and intercept of these lines will fluctuate too strongly to be useful. In fact, many Monte Carlo data are not precise enough to estimate three parameters simultaneously with reasonable accuracy, as in Eq. (A.2) (constant, exponent θ, and factor of proportionality). Fortunately, for percolation, the critical point p_c can be determined well by the iteration method of Section 2.7, independent of the data for any divergent quantity or generating function. Thus for medium-quality data we can take the percolation threshold from that method, or from the literature; then only two parameters are needed to describe the leading asymptotic behaviour of many quantities.

For example, we may look at the mean cluster size $S(p) \propto |p - p_c|^{-\gamma}$ near the threshold, or at the cluster size distribution $n_s(p_c) \propto s^{-\tau}$ at the threshold for large s. Then with medium-quality data one plots S or n_s double-logarithmically versus $|p - p_c|$ or s, respectively. Fitting a straight line through the data point gives an estimate for the exponent γ or τ from the slope, whereas the intercept gives the factor of proportionality. If $p - p_c$ or s vary by at least an order of magnitude, without taking into account data far away from p_c or small clusters, respectively, then the exponent estimates may be correct within about 10 per cent.

This fit of a straight line may be done visually on log-log paper or it may be done electronically. Many hand calculators have built in programs to determine slope and intercept automatically. If you want to program this fit for yourself you will find the formulae in suitable manuals or textbooks (least-squares fit). They are also derived easily by assuming that (taking $y_i = \ln(S)$ as function of $x_i = \ln|p - p_c|$) the sum of the squared deviations, $\Sigma_i(y_i - ax_i - b)^2$, from the straight-line fit $y = ax + b$ is as small as possible. Differentiating this sum with respect to a and b, and setting the results equal to zero, determines the slope a (exponent) and the prefactor b (proportionality factor).

Each single datum point has a statistical error. A rough estimate of this statistical error can be made by repeating the experiment with another set of random numbers. The difference then is of the order of the statistical error. A more reliable estimate is to make not just two but N independent simulations. We denote the average of a quantity y over these N simulations by $\langle y \rangle$, at one fixed parameter x. Then the probable statistical error (which is not the standard deviation) for y is

$$\Delta y = \left[\frac{\langle y^2 \rangle - \langle y \rangle^2}{N-1}\right]^{1/2} \tag{A.3}$$

If we are interested in only one value for y we can write $\langle y \rangle \pm \Delta y$ as our esti-mate and error bar for the quantity y. (The denominator $N-1$ is needed to account for the fact that your error is smaller the harder you have worked, that is the larger N is. On the other hand, with only one measurement, i.e. for $N = 1$, one has $\langle y^2 \rangle = \langle y \rangle^2$ even though the error is not zero. Thus it would not be legitimate to divide by N only. In any case, N should be large for precision measurements, and then the difference between N and $N-1$ is unimportant.)

Unfortunately, to estimate asymptotic properties these statistical errors are not enough. Usually we do not want to know what the cluster numbers are for $s = 1000$, but how they behave for $s \to \infty$, knowing them only for, say, $10 < s < 100$. Even if the statistical errors are exactly zero (as they are in series data) we cannot extrapolate to infinity with zero error. The additional deviations due to these necessary extrapolations are called systematic errors. Textbooks often contain formulae showing how to estimate the statistical errors for the slope and intercept of a straight-line fit from the statistical errors Δy of the single datum points. But these formulae are misleading for our applications since they ignore the systematic error. From only two exact points, say n_s for $s = 1$ and $s = 2$, these formulae would predict zero error for exponent and proportionality factor in $n_s \propto s^{-\tau}$. But we know that we cannot predict the behaviour for very large s with zero error if we know it only for $s = 1$ and $s = 2$.

How can we get better estimates for the quantities of interest and their error bars? First we need high precision for each point. Then, if we plot, say, n_s, versus s double logarithmically, we will see curvature in our data even for rather large s. This curvature tells us that a simple power law is not sufficient to describe the data. Thus instead let us try to work with

$$n_s(p_c) \propto s^{-\tau}(1 + \text{const}/s^\Omega + \cdots) \tag{A.4}$$

with one correction-to-scaling term proportional to $s^{-\Omega}$. Analogous assump-tions can be made for, say, the mean cluster size $S(p)$ as function of $p - p_c$ (see review of Adler *et al.* in the book *Percolation Structures and Processes* cited in Chapter 1). Of course, there will be more than just this one correction term to the leading $s^{-\tau}$ variation, but in general present series or Monte Carlo data are not accurate enough to determine a second correction term reliably if nothing else is known about it. Near Eq. (51) we discussed how to analyse such data.

Whichever method you use, you should also vary the set of data points on which you fit the straight lines. Usually, your data for small clusters (or for p far away from p_c) are more accurate than those for large s (or p near

p_c). Unfortunately we need mainly data for large clusters. Omitting the largest and/or the smallest cluster size will slightly change your estimates and give you a better impression of the systematic errors in your analysis.

If you have data varying over a large range of parameters, like the distances in Fig. 32 as a function of time up to 10^7, you may also go back to the good old log-log plot, determine the slope of the curve as a function of one of the variables, and extrapolate this slope to the asymptotic regime by plotting this slope versus some suitable power (exponent Ω) of the selected variable.

Generally, in a careful analysis you should try all methods which seem suitable. The spread of the results may give you a better impression of the error bars for your estimate than any one single method.

Unfortunately, even that is not enough. If you have found a suitable fit for, say, $n_s(p_c)$ versus s, you may still get something wrong since you had to employ finite systems. Sometimes (see Section 4.1) the effects of the finite system size can be used to determine some exponents, or can be taken into account by finite-size scaling. But if most of your data are made at one system size, and fewer runs are made at a drastically different system size, then a much simpler comparison of the two results tells you what order of magnitude the finite-size effects are. Quite often it also happens that programming errors give an influence which vanishes for system size going to infinity. In addition, you may get some systematic errors from random number generators which are not really random; trying different generators may, or may not, help.

In summary, it is difficult but possible to extract exponents from high-quality data with an accuracy of the order of 1 per cent. It is much easier to estimate errors of that order by using some standard formula ignoring systematic deviations, but then it is likely that later one will regret having published such an overly optimistic error bar. Our best advice is to try different methods and use some fantasy; different problems and different data qualities require different methods of analysis.

A.3. COMPUTERIZED CLUSTER COUNTING

The preceding sections of this appendix have discussed how to analyse cluster numbers. Redner's paper (see Mertens (1990) cited in Chapter 2) gives a short program for counting exactly the number of different cluster configurations for a given size. But how do we count clusters in a Monte Carlo sample of a large lattice? And how do we check whether a cluster connects top and bottom in such a sample? If one tries to do that visually one will presumably make some errors in a large lattice. We now explain how to teach a computer to do that work for us. We restrict ourselves to the algorithm of Hoshen and Kopelman (1976) since that allows the simulation of large lattices without having to store the whole lattice. Leath's entirely different algorithm, which involves letting one cluster grow, has already been mentioned in Chapter 2.

In general, simulations of a lattice are faster if one numbers all sites consecutively with one index, and not with d indices in d dimensions. For example, a site on a 10×10 square lattice does not need to be determined by two indices i and j between 1 and 10 each; we can also number them with one index K between 1 and 100. Then the first line has indices $1, 2, \ldots, 10$, the second line is numbered 11–20, and so on, until the last line with K from 91 to 100. The four nearest neighbours of a site K have the indices $K-1, K+1, K-L$ and $K+L$, on an $L \times L$ square lattice, thus the left neighbour of the leftmost site 21 of the third line is taken as site 20, the rightmost site of the second line ('helical boundary conditions'). For a simple cubic lattice the fifth and sixth neighbour are $K-L^2$ and $K+L^2$; for a triangular lattice, the are $K-L+1$ and $K+L-1$. However, these general methods are not even needed to count clusters on a square or triangular lattice, since then one line only has to be stored, as will now be explained.

What we would like to have is an algorithm which gives all sites within the same cluster the same label, and gives different labels to sites belonging to different clusters. Then the top of a sample is connected to its bottom if the same label appears in both the top and the bottom line or plane. And by counting how many sites have the same label we get the cluster size s.

Unfortunately, life is more difficult than our dreams. Let us look at the example of Fig. 39 and analyse it in the same way as you read this book: from left to right within each line, and then from the top line to the bottom line. We will give the first occupied site in the left upper corner the label 1; its neighbour to the right is empty and needs no label; then follows another occupied site, labelled by a 2, another empty site, and finally an occupied site labelled by a 3. The next line starts with an occupied site which is a neighbour to the occupied site labelled by a 1 in the first row. Thus we label this site by a 1, too. The next site is empty; the third site is labelled by a 2 since it is directly below the occupied site labelled 2 in the first row. The fourth site is neighbour to the third one and thus also labelled by a 2. Thus when we are looking at the fifth site our labels so far are:

$$
\begin{array}{ccccc}
1 & 0 & 2 & 0 & 3 \\
1 & 0 & 2 & 2 & ?
\end{array}
$$

where we marked the empty sites by zeros. What label do we choose for the fifth site denoted here by a question mark? Its lefthand neighbour says it is a 2 whereas its top neighbour claims it as its neighbour with a 3 label. In this

Fig. 39. Illustration of a 3×5 square lattice with 11 occupied sites, to be analysed by the Hoshen–Kopelman algorithm.

situation similar to that in the Fashoda swamps in 1898, one needs an *entente cordiale*. In reality all sites labelled 3 and all sites labelled 2 belong to one joint cluster which we label 2 in order to keep the label numbers as small as possible.

Having agreed on that common label at the question mark site, do we now have to go back to the beginning of the whole lattice to relabel all 3s into 2s? For small lattices we could do that and thus solve all these label conflicts by starting all over. But for large lattices that would make the computing time prohibitively large. We would like to have a computing time proportional to the number of lattice sites, and not to the square of that number, for large lattices.

Hoshen and Kopelman found a way to avoid this tedious relabelling. They follow some good old Western movies from Hollywood by dividing the whole set of labels into the good ones and the bad ones. The good ones are those which characterize different clusters. The bad ones are those which indicate that what first looked like a new cluster later turned out to be part of an old cluster. To distinguish easily between the good and the bad guys, we cannot distribute white and black hats to them. Thus instead we introduced an additional array, the labels of labels, and denote it as N. A good label, say M, is characterized by $N(M) = M$ whereas the label N of a bad label is taken as the label to which that bad label turned out to be connected.

Thus in our case, before we came to the question mark, we regarded our three labels as good and had $N(1) = 1$ when we found the first occupied site labelled by 1, took $N(2) = 2$ when we came to the second label 2, and set $N(3) = 3$ when the third label occurred for the first time. No changes were made in the array N when further sites with labels 1 and 2 were observed. Now we are at the site with the question mark. We give that site the label 2; note that 3 turned out to be a bad label since its sites are connected with sites of label 2, and thus take $N(3) = 2$.

Now we go to the third line of our above lattice. The first site is occupied and connected to the top site; thus we label it by 1 and leave $N(1)$ unchanged. The second site is labelled by a 1, since it too is connected to its left neighbour. At the third site we have a label conflict again since labels 1 (to the left) and 2 (to the top) turn out to belong to the same cluster. The good label is the smaller one, that means 1, and the bad label is the larger one, that means 2. Thus with $N(2) = 1$, we have marked that sites with label 2 belong to the same cluster as sites with label 1. Now our labels are

$$
\begin{array}{ccccc}
1 & 0 & 2 & 0 & 3 \\
1 & 0 & 2 & 2 & 2 \\
1 & 1 & 1 & 1 & 0
\end{array}
$$

$$N(1) = 1 \qquad N(2) = 1 \qquad N(3) = 2$$

In this way we can go through the whole lattice once and store the connections found later in the 'label of labels' array N. Once finished, for any site of the lattice we find its good label by the following classification: if the label of that site is M then check first if $N(M) = M$. If yes, the label is good and we go to the next site; otherwise the label is bad and equals, say, $N(M) = M'$. Now we check whether M' is a good label, $N(M') = M'$, or a bad one, $N(M') = M''$. In the first case the number M' is the root of our label tree and classifies the given site; in the other case we check if $N(M'')$, and so on, until we come to a good label with $N(\text{label}) = \text{label}$. With $M = 3$, in our case we have $N(3) = 2$ and thus $M' = 2$; then $N(2) = 1$ and thus $M'' = 1$. Since $N(1) = 1$, M'' is a good label whereas M and M' are bad.

A simple FORTRAN subroutine like

```
      FUNCTION KLASS (M)
      DIMENSION N(25000)
      COMMON N
1     MS = M
      M = N(M)
      IF(MS.NE.M) GO TO 1
      KLASS = M
      RETURN
      END
```

accomplishes the above classification: for an occupied site with original label M it searches for the good label at the root of the label tree and calls that good label KLASS.

It is practical when going through the lattice not to shift all classifications to the end of the calculation, although every site could get its good label with a second sweep through the lattice. Instead, before we assign a label to the newly investigated site, we may reclassify its two neighbours (in a square lattice), which got labels before, by the function KLASS. After the two currently good labels of these two neighbours have been found, the new site gets the smaller one as its own label, and the larger label is classified as bad and as connected to the smaller label if the site was occupied. This reclassification of the neighbours has the effect that after finishing one line (or plane in three dimensions) we can already tell if at least one occupied site is connected to the top line or plane. For if none of the labels of the line (or plane) just completed agrees with a label occurring in the first line (or plane) of the lattice, then the first line is not connected to the line under current investigation and thus also not to the bottom line. The investigation can then be stopped if we only want to know if top and bottom are connected.

Minor remarks: since in the case of a label conflict we take the smaller one, it is practical to take a very large number, like MAX with $N(\text{MAX}) = \text{MAX}$, as the label of empty sites. Then by simply looking at the smallest of all labels assigned to the previously investigated and reclassified neighbours we get the new label; we do not have to distinguish carefully between occupied and empty sites. Also, the left neighbour needs no reclassification since its label was set in the immediately preceding step and did not have the time to become bad. Finally, time-consuming IF conditions are avoided if we add an empty column to the left boundary of our square, and perhaps also an empty line to the top boundary.

Often one wants not just to see whether a lattice percolates but how many clusters of what size it contains. Then a modification is necessary. Now the bad labels have negative N(label), the good ones have positive N(label). The bad label's N equals minus the label to which it is connected (instead of plus that value in the connectivity check above), and the good label has an N(label) equal to the size of the cluster to which it belongs at present. Thus when a label conflict occurs, we select again as the new label the minimum of the (reclassified) labels of the previously investigated neighbours. But for N of this label we take the sum of the N of the previously separated clusters, plus unity for the site which we have just added. The classification subroutine now reads

```
      FUNCTION KLASS(M)
      DIMENSION N(25000)
      COMMON N
      MS = N(M)
      IF(MS.LT.0) GO TO 1
      KLASS = M
      RETURN
1     KLASS = - MS
      MS = N(KLASS)
      IF(MS.LT.0) GOTO 1
      N(M) = - KLASS
      RETURN
      END
```

(Some computers might complain that this is not really a function but a subroutine since we take the opportunity in the last statement to let the label $N(M)$ of the old label M point from now on to the just found good label KLASS.)

Computer time is saved if the classification routine is not regarded as a separate function KLASS but written into the main program. We then work in the simple cubic or triangular lattice with three previously investigated neighbours ('top', 'back' and 'left') and have to reclassify only two of them (top and back) if the left neighbour was dealt with in the previous step. We now list a complete FORTRAN program counting clusters on an $L \times L$ triangular lattice at various concentrations p, with $L = 500$. This half-minute test run was made on a CDC Cyber 76 computer; an IBM 3081 may take about a minute. Even vector computers do not run much faster.

```
        PROGRAM PERC(OUTPUT,TAPE6=OUTPUT)
        DIMENSION LEVEL(501),N(25000),NS(18)
        LOGICAL TOP,LEFT,BACK
C       TRIANGULAR SITE L*L PERCOLATION AT
C         CONCENTRATION P
        CALL RANSET(1)
        L=500
        LP1=L+1
        LARGE=1024
        ALOG2=1.0000001/ALOG(2.0)
        MAX=25000
        MAX3=MAX*3
        N(MAX)=MAX
        DO 7 IP=1,25
        P=0·37+0·01*IP
        DO 2 I=1,18
    2   NS(I)=0
        INDEX=INF=0
        CHI=0.
        DO 1 I=1,LP1
    1   LEVEL(I)=MAX
C       NOW ALL INITIAL CONDITIONS ARE SET
C       THE FIRST LINE AND LEFTMOST COLUMN IS EMPTY
        DO 3 K=2,LP1
        IF(INDEX.GT.24700)GOTO 7
C       DANGER OF MEMORY OVERFLOW, STOP WORKING
        MOLD=MAX
        DO 3 I=2,LP1
        LBACK=MBACK=MOLD
        MOLD=LEVEL(I)
        IF(RANF(I).GT.P)GOTO 9
C       JUMP TO 9 IF NEW SITE IS EMPTY
        MLEFT=LEVEL(I-1)
        MTOP=LTOP=LEVEL(I)
        IF(MLEFT+MTOP+MBACK.EQ.MAX3)GOTO 4
C       JUMP TO 4 IF ALL THREE NEIGHBOURS ARE EMPTY
        LEFT=MLEFT.LT.MAX
        TOP =MTOP .LT.MAX
        BACK=MBACK.LT.MAX
C       FIRST HOSHEN-KOPELMAN CLASSIFICATION OF TOP
C         NEIGHBOUR
        IF(.NOT.TOP.OR.N(LTOP).GE.0) GOTO 12
        MS=N(LTOP)
   13   MTOP=-MS
```

```
C       THIS IS THE FUNCTION KLASS WRITTEN INTO MAIN
C         PROGRAM
        MS = N(MTOP)
        IF(MS.LT.0) GOTO 13
        N(LTOP) = - MTOP
C       NOW COMES THE BACK NEIGHBOUR (LEFT OF TOP)
    12  IF(.NOT.BACK.OR.N(LBACK).GE.0) GOTO 11
        MS = N(LBACK)
    14  MBACK = - MS
        MS = N(MBACK)
        IF(MS.LT.0) GOTO 14
        N(LBACK) = - MBACK
C       LEFT NEIGHBOUR NEEDS NO RECLASSIFICATION
    11  LEVEL(I) = MNEW = MIN0(MTOP,MBACK,MLEFT)
C       MIN0 GIVES SMALLEST OF SEVERAL INTEGERS
        ICI = 1
        IF(TOP)ICI = ICI + N(MTOP)
        IF(LEFT.AND.MTOP.NE.MLEFT) ICI = ICI + N(MLEFT)
        IF(BACK.AND.MBACK.NE.MLEFT.AND.MBACK.NE.MTOP)
     1    ICI = ICI + N(MBACK)
        N(MNEW) = ICI
C       ICI IS THE SIZE OF THE CLUSTER AT THIS STAGE
        IF(TOP .AND.MTOP .NE.MNEW)N(MTOP ) = - MNEW
        IF(LEFT.AND.MLEFT.NE.MNEW)N(MLEFT) = - MNEW
        IF(BACK.AND.MBACK.NE.MNEW)N(MBACK) = - MNEW
        GOTO 3
     4  LEVEL(I) = INDEX = INDEX + 1
C       START OF NEW CLUSTER
        N(INDEX) = 1
        GOTO 3
     9  LEVEL(I) = MAX
     3  CONTINUE
C       NOW FINAL ANALYSIS
        IF(INDEX.EQ.0) GOTO 5
        DO 6 IS = 1,INDEX
        NIS = N(IS)
C       IF NIS < 0 IT IS A BAD LABEL AND SHOULD BE IGNORED
        IF(NIS.LT.0) GOTO 6
C       NIS IS THE NUMBER OF CLUSTERS CONTAINING S SITES
C         EACH
C       INF IS THE SIZE OF THE LARGEST CLUSTER
C       CHI IS THE SECOND MOMENT, RELATED TO MEAN
C         CLUSTER SIZE
        IF(INF.LT.NIS)INF = NIS
        FNIS = NIS
```

```
          CHI = CHI + FNIS*FNIS
          IF(NIS.GE.LARGE) WRITE(6,97) NIS
    C     LARGE CLUSTERS ARE PRINTED OUT SEPARATELY,
    C        SMALLER
    C     ONES ARE PUT TOGETHER IN BINS FROM 2**(I – 1) TO
    C        2**I – 1
          NIS = ALOG(FNIS)*ALOG2 + 1.
          NS(NIS) = NS(NIS) + 1
    6     CONTINUE
          CHI = (CHI – FLOATING(INF)**2)/ (L*L)
    5     WRITE(6,96) NS
   96     FORMAT(" NS: ",4I8,4I6,/,10I6)
   97     FORMAT(" CLUSTER OF SIZE ",I12)
   98     FORMAT(F10·5,3I10,F20·4,/)
    7     WRITE(6,98)P,L,INDEX,INF,CHI
          STOP
          END
```

We see from this computer program that the original labels, called LEVEL here, do not require an $L \times L$ array; only one line of length L has to be stored in two dimensions, and one plane in three. However, the array N which stores the labels of labels has to have a size proportional to the number of sites in the system. We took its size here as 25 000 with $L = 500$ being our maximum size. Thus N requires about one-tenth of the system size in memory. Actually, every occupied site with all previously investigated neighbours being empty increases our index for the labels by unity and thus requires one more memory element in the array N. The probability for this event to happen is $p(1 - p)^3$ in the triangular or simple cubic lattice and agrees reasonably, if multiplied by L^2, with the final value of INDEX printed out below. (For bond percolation at very small concentration p, the array N must be about as large as the whole lattice, if one takes no precaution to throw out isolated sites.)

Periodic boundary conditions often diminish, and sometimes enhance, the influence of the lattice boundaries; in any case they make the program more complicated. Computer memory can be saved, or larger systems be simulated, if labels of labels which are no longer used in the array N are recycled like used paper. This trick is described in Chapter 8 of Binder's 1987 book on Monte Carlo methods, listed after Chapter 2, with a complete program given there. Thus we will not go into these details and now look at the output from the simpler program listed above.

In the following computer output from a CDC Cyber 76 we give the results for many different concentrations p. In every example, the first two lines give the cluster numbers in bins of exponentially increasing size: $s = 1$, $s = 2$ and 3, s from 4 to 7, then from 8 to 15, etc., as mentioned in Section 2.7. Then we list five other interesting quantities: the concentration p, the lattice size L, the maximum index needed (a number which must be smaller than the memory size reserved for the array N), the size of the largest cluster, and the second moment of the cluster size distribution (which enters the mean cluster size). Also, we list the precise size of all clusters containing at least LARGE = 1024 sites, in order to study the second or third largest cluster too. Thus it already becomes transparent from the output that lots of large clusters only occur close to the percolation threshold. In Section 2.7 we described how we analysed these low-quality Monte Carlo data.

15.52.50   00031·711 USR.   31·405 CP SECONDS EXECUTION TIME.

NS: 5493   3851   2640   1720   858   416   112   14
       0   0   0   0   0   0   0   0   0   0
       ·38000   500   22838   211   10·6030

NS: 4994   3458   2331   1595   878   447   149   26
       0   0   0   0   0   0   0   0   0   0
       ·39000   500   22094   218   13·3170

NS: 4754   3326   2162   1461   876   450   185   35
       2   0   0   0   0   0   0   0   0   0
       ·40000   500   21838   295   16·0111

NS: 4458   2969   1973   1309   846   468   197   52
       5   1   0   0   0   0   0   0   0   0
       ·41000   500   21225   614   20·1037

NS: 4124   2738   1810   1261   736   462   216   72
       9   1   0   0   0   0   0   0   0   0
       ·42000   500   20725   540   26·4825

NS: 3680   2447   1657   1033   671   414   200   105
       24   2   0   0   0   0   0   0   0   0
       ·43000   500   20009   664   38·0001

CLUSTER OF SIZE   1132
NS: 3436   2292   1349   963   607   328   190   109
       37   9   1   0   0   0   0   0   0   0
       ·4400   500   19381   1132   57·3412

NS: 3138   2010   1289   816   539   338   217   120
       45   10   0   0   0   0   0   0   0   0
       ·45000   500   18703   870   62·4797

CLUSTER OF SIZE   2205
CLUSTER OF SIZE   1808
CLUSTER OF SIZE   1063
NS: 2911   1856   1133   724   468   281   169   95
       51   26   2   1   0   0   0   0   0   0
       ·46000   500   18201   2205   107·4869

CLUSTER OF SIZE   1493
CLUSTER OF SIZE   2225
CLUSTER OF SIZE   1045
CLUSTER OF SIZE   2498
CLUSTER OF SIZE   2160
CLUSTER OF SIZE   2267
CLUSTER OF SIZE   1498
CLUSTER OF SIZE   1038
CLUSTER OF SIZE   1328
CLUSTER OF SIZE   1030

NS:	2541	1671	985	602	348	220	134	80
50	31	6	4	0	0	0	0	0	0

·47000   500   17574   2498   206·8616

CLUSTER OF SIZE   1536
CLUSTER OF SIZE   2335
CLUSTER OF SIZE   3054
CLUSTER OF SIZE   1552
CLUSTER OF SIZE   1252
CLUSTER OF SIZE   4043
CLUSTER OF SIZE   1209
CLUSTER OF SIZE   1479
CLUSTER OF SIZE   1933
CLUSTER OF SIZE   2240
CLUSTER OF SIZE   2204
CLUSTER OF SIZE   1479
CLUSTER OF SIZE   6700
CLUSTER OF SIZE   2149
CLUSTER OF SIZE   3078
CLUSTER OF SIZE   2511
CLUSTER OF SIZE   1303
CLUSTER OF SIZE   1709
CLUSTER OF SIZE   1625
CLUSTER OF SIZE   1112
CLUSTER OF SIZE   1538
CLUSTER OF SIZE   1286

NS:	2423	1420	876	530	280	182	91	51
36	17	13	8	1	0	0	0	0	0

·48000   500   16945   6700   425·9939

CLUSTER OF SIZE   9668
CLUSTER OF SIZE   2541
CLUSTER OF SIZE   1310
CLUSTER OF SIZE   1317
CLUSTER OF SIZE   1491
CLUSTER OF SIZE   2200

```
CLUSTER OF SIZE    2550
CLUSTER OF SIZE    3695
CLUSTER OF SIZE    1190
CLUSTER OF SIZE    2749
CLUSTER OF SIZE   10429
CLUSTER OF SIZE    2199
CLUSTER OF SIZE    2506
CLUSTER OF SIZE   15059
CLUSTER OF SIZE    1272
CLUSTER OF SIZE    1075
CLUSTER OF SIZE    2073
```

NS:	2281	1347	797	456	250	137	77	48
22	14	3	8	0	3	0	0	0
·49000		500	16428	15059			1117·5676	

```
CLUSTER OF SIZE   19937
CLUSTER OF SIZE   16111
CLUSTER OF SIZE    3550
CLUSTER OF SIZE    4311
CLUSTER OF SIZE   24895
CLUSTER OF SIZE    1147
CLUSTER OF SIZE    5195
CLUSTER OF SIZE    1876
CLUSTER OF SIZE    2170
CLUSTER OF SIZE    1307
```

NS:	1992	1173	699	367	195	89	58	32
10	13	3	2	2	1	2	0	0
·50000		500	15729	24895			2947·0190	

```
CLUSTER OF SIZE   76513
CLUSTER OF SIZE    1964
CLUSTER OF SIZE    1123
CLUSTER OF SIZE    1875
CLUSTER OF SIZE    2339
CLUSTER OF SIZE   10860
CLUSTER OF SIZE    1453
```

NS:	1868	1041	549	312	143	84	39	19
14	3	4	1	0	1	0	0	1	0
·51000		500	15306	76513			554·8937	

```
CLUSTER OF SIZE    2272
CLUSTER OF SIZE   99041
CLUSTER OF SIZE    1293
CLUSTER OF SIZE    3145
```

NS:	1667	951	518	254	149	54	40	13
3	3	1	2	0	0	0	0	1	0
·52000		500	14754	99041			77·0606	

CLUSTER OF SIZE 118338

NS:	1456	818	401	218	89	33	13	7
2	0	0	0	0	0	0	0	1	0

·53000   500   14075   118338   3·0231

CLUSTER OF SIZE 125000

NS:	1355	663	355	144	66	18	10	2
0	1	0	0	0	0	0	0	1	0

·54000   500   13405   125000   2·0732

CLUSTER OF SIZE 128563

NS:	1158	625	311	119	62	23	5	1
0	0	00	0	0	0	0	1	0

·55000   500   12959   128563   ·6700

CLUSTER OF SIZE 133636

NS:	1065	526	229	101	35	11	0	1
0	0	0	0	0	0	0	0	0	1

·56000   500   12236   133636   ·3457

CLUSTER OF SIZE 137726

NS:	916	436	186	87	28	8	1	0
0	0	0	0	0	0	0	0	0	1

·57000   500   11473   137726   ·2070

CLUSTER OF SIZE 141958

NS:	814	336	158	60	12	1	1	0
0	0	0	0	0	0	0	0	0	1

·58000   500   10872   141958   ·1042

CLUSTER OF SIZE 143765

NS:	774	325	136	47	13	5	0	0
0	0	0	0	0	0	0	0	0	1

·59000   500   10491   143765   ·1177

CLUSTER OF SIZE 147201

NS:	709	266	120	26	8	1	1	0
0	0	0	0	0	0	0	0	0	1

·60000   500   10012   147201   ·0963

CLUSTER OF SIZE 150802

NS:	576	241	75	14	7	1	1	0
0	0	0	0	0	0	0	0	0	1

·61000   500   9037   150802   ·0617

CLUSTER OF SIZE 153402

NS:	501	217	75	24	4	1	0	0
0	0	0	0	0	0	0	0	0	1

·62000   500   8659   153402   ·0461

Output of this type was used by Margolina *et al.* (1983) (cited in Chapter 2) to analyse lattices with up to 10^{10} sites, using only one megabyte of a CDC Cyber 76. Clearly that would have been impossible without the Hoshen–Kopelman algorithm with its elegant classification routine and the recycling of labels of labels no longer needed in the array N. With additional tricks, Rapaport (1985) (as cited in Chapter 2) simulated at $160\,000 \times 160\,000$ square lattice; today a workstation can repeat his world record on a weekend. In spring 1991, Rapaport simulated a $640\,000 \times 640\,000$ lattice.

Readers interested in diffusion on disordered lattices will find FORTRAN programs, including those for CDC Cyber 205 vector computers, in the paper by Pandey *et al.* (1984) cited in Chapter 6.

FURTHER READING

Hoshen, J. and Kopelman, R., *Phys. Rev. B*, **14**, 3428 (1976).

APPENDIX B

Dimension-Dependent Approximations

B.1. UPPER CRITICAL DIMENSION

We start with a discussion of *polymer chains*, or *self-avoiding* walks, and *branching polymers*, or *lattice animals*, both in a dilute solution with no interaction to other chains or polymers. As we mentioned in Section 5.3, a polymer chain is a random chain of monomers (except for their connections to their nearest neighbours). The chain with s monomers behaves as a random walk (diffusion) of s time steps. Therefore, its radius R_s varies as $\sqrt{s}$. For polymers with random branchings, the lattice animal description may be a good model, and there the Bethe lattice approximation gives $R_s \propto s^{1/4}$ (Table 2) just as for percolation clusters at p_c. We call the radius in these simple limits R_{s0}, thus $s \propto R_{s0}^{D_0}$ with the fractal dimensions $D_0 = 2$ and 4 for chains and branched animals respectively.

In reality, we have *interactions*, within the same chain or polymer, neglected in these approximations, which may lead to a lower fractal dimension D and a larger radius $R_s \propto s^{1/D}$. The interactions become important if different pieces of the polymer (or the animal) have a finite probability of getting close to each other. The density of the non-interacting polymer, in a volume of linear size R_{s0}, is $\rho \propto R_{s0}^{D_0 - d}$. Therefore, the probability per unit volume that the polymer goes twice through that unit volume is proportional to ρ^2, and the total number of such intersections is of order $R_{s0}^d \rho^2 \propto R_{s0}^{2D_0 - d}$. These intersections thus become negligibly rare for large R_{s0} and $d > d_u = 2D_0$, where d_u is the 'upper critical dimension'. This dimension is $d_u = 4$ for polymer chains and $d_u = 8$ for lattice animals. At $d > d_u$ we can thus describe the polymer as non-interacting, and $D = D_0$.

For percolation clusters at p_c, we saw that for the Bethe lattice (and therefore for high dimensions) $D_0 = 4$, so that the cluster looks somewhat like a branched polymer, with a few large loops. At high dimensions, the cluster backbone behaves like a self-avoiding walk, with $D_{0B} = 2$. To create a blob, we need the backbone to intersect with some otherwise dangling bond. The densities of these two structures are $\rho_0 \propto R_s^{D_0 - d}$ and $\rho_{0B} \propto R_s^{D_{0B} - d}$, and thus the number of their intersections is proportional to $R_s^d \rho_0 \rho_{0B} \propto R_s^{D_0 + D_{0B} - d}$.

Therefore, the upper critical dimension, above which this picture is self-consistent, is $d_u = D_0 + D_{0B} = 6$. See also the discussion after Eq. (53).

Another way to identify $d_u = 6$ is to note that the Bethe lattice results $sn_s \propto s^{-3/2}$ and $D = 4$ imply that the number of spanning clusters of size ξ is of order $\xi^d (\xi^4)^{-3/2} = \xi^{d-6}$. Therefore, the number of spanning clusters is larger than one for $d > 6$. Since the mass and the conductance connecting the edges of the sample arise from *all* the spanning clusters, they contain extra factors of ξ^{d-6} for $d > 6$ (see discussion at end of Section 5.3).

B.2. FLORY APPROXIMATION

For $d < d_u$, the density of self-interactions grows with R_{s0}, and this yields a new behaviour. This new behaviour is not trivial and (unlike the results for $d > d_u$) there exist only very few exact results. The *Flory approximation* is based on the competition between the elastic energy and the self-repulsion. The former tries to keep R close to R_{s0}, as a spring tries to keep its length, and is proportional to $[(R_s - R_{s0})/R_{s0}]^2$. The latter is proportional to the number of self-intersections, which is again of order $R_s^d \rho^2$. However, now $\rho \propto s/R_s^d$, and hence the interaction energy is proportional to s^2/R_s^d. Thus the total energy is

$$E_s \propto \frac{(R_s - R_{s0})^2}{R_{s0}^2} + \frac{As^2}{R_s^d} \tag{B.1}$$

For $d > d_u = 2D_0$, if the proportionality factor A is independent of s, then E_s is minimized by $R_s = R_{s0} \propto s^{1/D_0}$ (for large s) and we recover the non-interacting behaviour. For $d < d_u$, the minimum is found via $(dE_s/dR_s) = 0$, just as for a string under an external force. This yields (for large s)

$$R_s^{d+2} = \frac{1}{2} dA s^{2+2/D_0}$$

or $s \propto R_s^D$ with

$$D = \frac{2+d}{2+2/D_0}$$

Thus we have

$$D = (2+d)/3 \qquad \text{for chains} \tag{B.2a}$$

and

$$D = 2(2+d)/5 \qquad \text{for animals} \tag{B.2b}$$

Equation (B.2a) for these 'self-avoiding walks' turns out to be exact for $d = 1, 2$, and 4 and correct within 2 per cent in three dimensions. The animal D of Eq. (B.2b) is exact for $d = 3, 4$ and 8, compatible with numerical data for $d = 5, 6$, and 7, and too large by $0 \cdot 04$ in two dimensions.

Although animals and percolation clusters have the same fractal dimensions $D_0 = 4$ at high dimensions, they have different upper critical dimensions. Therefore, it is clear that we cannot describe both by the same Eq. (B.1). In order to overcome this difficulty, de Gennes (1980) introduced an *ad hoc* screening factor, which involves a decay of the prefactor A with large s as $A \propto s^{-x}$. The exponent x can then be tuned so as to reproduce $d_u = 6$. Using the modified Eq. (B.1), the second term becomes negligible with $R_s \propto R_{s0} \propto s^{1/4}$ if $d > d_u = 4(2 - x)$. Hence, for percolation we need $x = 1/2$. For $d < d_u = 6$, minimization now yields

$$D = \frac{2 + d}{2 - x + 2/D_0} = (2 + d)/2 \qquad \text{for percolation at } p_c \qquad (B.2c)$$

This approximate formula is accurate within 1 per cent for $d = 3, 4$ and 5, and is 5 per cent too high for $d = 2$.

B.3. ε-EXPANSION

In the approximate Flory formulae, d can be treated as a free *continuous parameter* between 1 and d_u, and it need not be restricted to integer values. Such generalizations into non-integer dimensions have proved to be very useful in critical phenomena, starting with Wilson and Fisher who calculated exactly a few coefficients in expansions of critical exponents in powers of the parameter $\varepsilon = d_u - d$. Such calculations, based on renormalization group treatments of appropriate *field theories*, have since been done for most of the exponents discussed in this book. Since details are too complicated for this introductory text, we list only the leading terms in Table 2. Although valid only asymptotically close to d_u, extrapolation from $d_u = 6$ even down to $d = 2$ for percolation properties near or at p_c often gives surprisingly reasonable results. Note also that the ε-expansions for D near d_u do not agree with those for the corresponding Flory formulae, reflecting the approximate nature of the latter. Also, the ε-expansions for μ, β and ν do not agree with the Alexander–Orbach conjecture $d_s = 4/3$. Generally, new scaling hypotheses are more trustworthy if they agree to all known orders with the ε-expansion.

FURTHER READING

Flory approximations
de Gennes, P.G., *Scaling Concepts in Polymer Physics* (Ithaca, NY: Cornell University Press, 1979).
de Gennes, P.G., *Comptes Rend. Acad. Sci. Paris*, **291**, 17 (1980).
Flory, P.J., *Statistical Mechanics of Chain Molecules*, (New York: Interscience, 1969).
Isaacson, J. and Lubensky, T.C., *J. Phys. (Paris)*, **41**, L469 (1980).

ε-Expansions
Aharony, A., *Phys. Rev. B*, **22**, 400 (1980).
Harris, A.B., *Phys. Rev. B.*, **35**, 5056 (1987).
Harris, A.B. and Lubensky, T.C., *Phys. Rev. B*, **35**, 6964 (1987).
Park, Y., Harris, A.B. and Lubensky, T.C., *Phys. Rev. B*, **35**, 5048 (1987)
Stephen, M.J., *Phys. Rev. B*, **17**, 4444 (1978).

Exercises

CHAPTER 1

1.1 (a) Program a computer to print a square lattice with occupation probability p. (b) Use this program to produce pictures like Fig. 2, identify (manually) the largest cluster on each of them, and draw its linear size (largest end-to-end distance) and the number of its sites as a function of p. What can you conclude from these graphs? (c) At $p = p_c = 0\cdot59$, count the number M of sites on the largest cluster in boxes of size $3 \times 3, 5 \times 5, 7 \times 7, ..., L \times L$ around a site that belongs to the cluster and plot log M versus log L. What do you conclude from these plots?

1.2 (a) Prove that $R^2 = t$ for diffusion in an ordered lattice, i.e. $p = 1$, at any dimension. (b) Write a computer program for a random walk in one dimension, run it many times, and draw a histogram for the number of times the walker reaches a distance r from the origin after t time steps. Compare the resulting graph with $\exp(-r^2/2t)$.

CHAPTER 2

2.1 (a) Calculate the product of the number of nearest neighbours, z, and the bond percolation threshold for each of the lattices listed in Table 1 in two and three dimensions. What approximate rule do you identify by looking at these products? (b) Imagine a sphere (or circle), with a diameter equal to the distance between nearest neighbours, centred around each lattice site. For site percolation, each occupied site implies filling the volume (area) of the corresponding sphere. Calculate the volume (area) fraction of occupied spheres at the percolation threshold for each of the three-(two-)dimensional cases listed in Table 1. What is the resulting approximate rule? Based on this, can you conjecture a rule for the percolation threshold of spheres which are packed randomly in a box?

2.2 Use the exact solution in one dimension to show that the kth moment of the cluster size distribution, $M_k = \Sigma_s s^k n_s$, diverges as $\Gamma_k (1 - p)^{1-k}$, and calculate explicitly the amplitudes Γ_k.

2.3 (a) For one dimension, calculate the site cluster numbers for the case that

173

sites are occupied with probability p and bonds are occupied with probability x. (b) Use the results to repeat Exercise 2.2. (c) Calculate the ratios $\Gamma_k \Gamma_l / \Gamma_m \Gamma_{k+l-m}$ for both cases. What do you conclude about these ratios?

2.4 Consider a general Bethe lattice, in which each site has z neighbours. (a) Prove that for site percolation, the strength of the infinite cluster is given by $P = p(1 - Q^z)$, where Q must obey the equation $Q = 1 - p + pQ^{z-1}$. Expand $Q^{z-1} = [1 - (1 - Q)]^{z-1}$ in a Taylor expansion up to quadratic order in $(1 - Q)$ in order to obtain the approximate solution $P = B(p - p_c)$, and find p_c and B. (b) Repeat the derivation of Eq. (17), and identify the amplitude Γ in $S = \Gamma / (p_c - p)$. (c) Find the proportionality factor a in the asymptotic relation $c = a(p - p_c)^2$ in Eq. (23). (d) Repeat the arguments which led to Eq. (23), but keep all the amplitudes. Show that the result can be written in the form of Eq. (36), so that q_0 and q_1 depend on z but the function $f(x)$ is universal (i.e. z-independent).

2.5 (a) For $z = 3$, show that Eq. (16) yields

$$P = B(p - p_c) + C(p - p_c)^2 + \cdots$$

and identify the coefficients B and C. (b) Use Eq. (16) to plot log P versus $\log(p - p_c)$. Why do you not find a straight line? What determines the range of $(p - p_c)$ where the line is straight?

2.6 Use Eq. (36), with $f(z) = e^{-z}$, to calculate P and S explicitly for $p > p_c$, with all the proportionality factors ('amplitudes'). Calculate also the amplitude for S below p_c, and prove that the ratio R does not depend on q_0 and q_1.

2.7 (a) Give the exact $n_s(p)$ for $s = 1, 2, 3, 4$, for both site and bond percolation, on the triangular lattice. (b) Use the results to obtain low concentration series for $S(p)$. (c) Use the ratio and the Dlog–Padé methods on these series, to estimate p_c and γ.

2.8 Use Eq. (24) to derive an expansion of P in powers of $q = 1 - p$ for $q \ll 1$. For site percolation on the square lattice, show that

$$P = 1 - q - q^4 + q^5 - 4q^6 - 4q^7 + \cdots$$

CHAPTER 3

3.1 (a) Calculate $R_s^2(p)$ for site percolation on the square lattice and $s = 1, 2, 3, 4$ exactly. (b) Use the results and Eq. (47b) to calculate a low-concentration series for ξ^2.

3.2 Given a very large finite cluster, of mass s, at p_c, one measures the mass which is connected to its centre of mass within a box of linear size L, $M(L, s)$. (a) Discuss the dependence of M on L and on s. Writing $M(L, s) = L^A m(L/s^B)$, identify the exponents A and B, and find the behaviour of the scaling function $m(x)$ for $x \ll 1$ and $x \gg 1$. (b) The mass M can also be studied as a function of L and R_s. Repeat the above discussion for $M(L, R_s)$.

CHAPTER 4

4.1 Use the one-dimensional result $\Pi = p^L$ to calculate p_{av} and Δ as functions of L. Use the plot of Δ versus p_{av} to 'find' p_c, and then the plot of $\log(p_c - p_{av})$ versus $\log L$ to find ν. For what range of L does this method work?

4.2 Calculate $S(L, p)$ for one-dimensional site percolation on a finite segment of L sites, with periodic boundary conditions. Discuss in detail the scaling properties of S, and the limits $L \ll \xi$ and $L \gg \xi$.

4.3 In a bond percolation renormalization group on the square lattice, the lattice is replaced by a new lattice, with bond of length $\sqrt{2}$, as shown by the lines in the figure on the left. In a truncated approximate treatment,

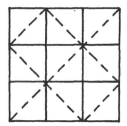

Exercise 4.3

the new bond concentration p' is calculated by checking connectivity within a single square, as drawn on the right. Show that the recursion relation is

$$p' = 2p^2 - p^4$$

and use it to find p_c and ν.

4.4 Do the same as above, for the honeycomb lattice (see figure). Show that

$$p' = 2p^3 - p^6$$

and find p_c and ν.

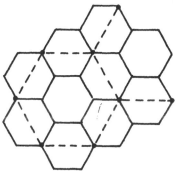

Exercise 4.4

4.5 Assume that $\Delta \propto b^{-1/\nu}(1 + C/b^{\Omega} + \cdots)$ for the width Δ in Eq. (69). How would you determine ν from Monte Carlo values of Δ?

CHAPTER 5

5.1 Find the values of L_1, L_2, L_3, L_4 and b which generalize the Mandelbrot–Given curve to yield good imitations of the critical percolation cluster in four and five dimensions, and calculate the corresponding values of $D_B, D_{\min}, D_{\max}, D_{SC}$ and $\tilde{\zeta}_R$. What are the corresponding values for $d \geqslant 6$?

5.2 Prove Eq. (95) for the multifractal exponents of the current distribution on the generalized Mandelbrot–Given curve.

5.3 There are two possible definitions of the backbone: (a) All the internal bonds which are connected by different paths to the two terminals, or the union of all the self-avoiding paths between the two terminals; (b) All the bonds which carry non-zero currents when a voltage is put between the two terminals. Numerical simulations show that the masses of the two backbones have the same fractal dimensions. Calculate these two fractal dimensions using the renormalization group scheme of Fig. 19 and comment on the results.

5.4 Use the bond renormalization scheme of Fig. 19 to calculate the multifractal exponents $y(q)$ for the current distribution on the critical percolation cluster in two dimensions. Use the results to evaluate $D_B, \tilde{\zeta}_R, y(2)$ and D_{SC}.

5.5 Use the renormalization schemes given in Exercises 4.3 and 4.4 to obtain estimates for the exponents $D_{SC}, D_{\min}, D_{\max}, D_B$ and $\tilde{\zeta}_R$.

5.6 (a) Two bonds are connected in parallel, and each has a conductance whose distribution is given by Eq. (98). Show that the distribution of the net conductance G is given by

$$f'(G) = \int_0^{\sigma_{\max}} d\sigma \, f(\sigma)f(G - \sigma) \propto G^{1-2w}$$

(b) When the two bonds are connected in series, with distribution functions $f_1(\sigma)$ and $f_2(\sigma)$, give arguments why for very small net conductance G one has

$$f'(G) \simeq f_1(G) + f_2(G)$$

(c) Use the above results to show that if $0 < w < 1$, then the net $f'(G)$ for the Mandelbrot–Given curve and for the renormalization scheme of Fig. 19 are dominated by singly connected bonds, i.e. $f'(G) = M_{SC}(b)f(G)$.

(d) If we rewrite Eq. (98) in the form $f(\sigma) \, d\sigma = (\sigma_0/\sigma)^w \, d(\sigma/\sigma_0)$, then show that $f'(G) \, dG = (\sigma_0'/G)^w \, d(G/\sigma_0')$, with $\sigma_0' = b^{1/(\nu(w-1))}\sigma_0$. Explain how this yields the results of Section 5.7.

CHAPTER 6

6.1 (a) An ant performs a random walk on a finite large cluster, of s sites, at p_c. Write down a scaling form for the root mean square distance R as function of t and s, and discuss in detail the limiting behaviour for short and long times. (b) For long times, one observes that $R = R'_s + Ae^{-t/\tau}$. Use the scaling results of part (a) to deduce how the relaxation time τ depends on s.

6.2 (a) An ant begins a random walk at the central site on a one-dimensional cluster of three sites. Prepare a table of $P_i(t)$ for 10 time steps, for both a 'blind' and a 'myopic' ant. What do you conclude on the stationary limits P_i(stationary) for these cases? (b) Assuming that $P_i(t) = P_i$(stationary) $+ A(-1)^t e^{-t/\tau}$, use Eq. (104) to find the relaxation times τ for these two cases.

6.3 (a) Prove that for branched polymers, where loops are not important,

$$d_s = \frac{2D}{D + D_{\min}} = \frac{2D_m}{D_m + 1}$$

where $D_m = D/D_{\min}$. Give D_m a geometrical interpretation, and explain why it may be considered more 'intrinsic' than D. (b) For walks on the backbone of branched polymers, prove that the fracton dimensionality is $d_{SB} = 1$.

6.4 Describe biased walks which count the accessible perimeter sites contained in E_1, E_2 and E_3.

6.5 For diffusion fronts, the mass on the front is a function of both L and ξ. Use scaling arguments to show that for $L \gg \xi$, this mass behaves as $L\xi^{D_h-1}$. How is this generalized to d-dimensional fronts?

CHAPTER 7

7.1 (a) At p_c, the low-temperature magnetization of a dilute magnet is given by

$$m \propto \Sigma s^{1-\tau} \tanh\left(\frac{sH}{kT}\right)$$

Use arguments like those leading to Eq. (32) to show that the singular dependence of m on H is of the form $m \propto H^{1/\delta}$, and prove that

$$\frac{1}{\delta} = \tau - 2 = \frac{\beta}{\beta + \gamma} = \frac{d - D}{D}$$

(b) For $p < p_c$, show that

$$m(p, H) = (p_c - p)^\beta f(Hs_\xi)$$

and discuss the behaviour of $f(x)$ for small and large x.

(c) 'Metastable' states and hysteresis are obtained in a magnet if for $p > p_c$ the magnetization m is oriented antiparallel to the field H. Determine the metastable magnetization from the cluster numbers in the dilute low-temperature Ising model. (Hint: Infinite cluster antiparallel to H.) Does this formula lead to a 'spinodal line', as many approximate theories of metastability do? (At spinodal lines, dm/dH diverges even far away from the Curie point.) Does your formula lead to a finite nucleation rate (Section 7.3), or have the metastable states an infinite lifetime?

7.2 Determine, from the known literature values of the Ising critical point for the square and simple cubic lattices, the bond probability in the Coniglio–Klein definition.

7.3 Determine from the known bond percolation thresholds the temperature at which the Coniglio–Klein droplets (without ghost spin correction) percolate at infinitely strong magnetic fields, for the square and simple cubic lattices.

7.4 Program the Ising model on the square lattice with the Swendsen–Wang technique, for zero magnetic field. (Hint: First learn the Hoshen–Kopelman technique explained in the Appendix A, Section A.3, to characterize percolation clusters.)

Index

A table of thresholds appears on p. 17.
A table of exponents appears on p. 52.

Printed in the United States
by Baker & Taylor Publisher Services